SUPERABRASIVES

Grinding and Machining With CBN and Diamond

S. F. Krar
Kelmar Associates

E. Ratterman
GE Superabrasives

GLENCOE/McGRAW-HILL
A Macmillan/McGraw-Hill Company

Westerville, Ohio   Mission Hills, California   Peoria, Illinois

Sponsoring Editor: Stephen M. Zollo
Editing Supervisor: Kelly Warsak
Design and Art Supervisor: Meri Shardin
Production Supervisor: Al Rihner

Interior Designer: LMD and Robin Hessel Hoffman
Cover Designer: Peri Zules

Cover photo: Setup for hard finishing of gears on a Kashifuji CNC machine, using a Borazon CBN tool. Background: Borazon CBNB 510 crystals *(Courtesy of General Electric Co.)*

Library of Congress Cataloging-in-Publication Data
Krar, Stephen F.
  Superabrasives: grinding and machining with CBN and diamond/
Stephen F. Krar, Ernest Ratterman.
    p. cm.
  ISBN 0-07-035587-8
  1. Abrasive—Congresses.   2. Grinding and polishing—Congresses.
3. Diamonds, Industrial—Congresses.   4. Boron nitride—Congresses.
I. Ratterman, Ernest. II. Title.
TJ1296.K73 1990
621.9′2—dc20   89-12841
                                                          CIP

**SUPERABRASIVES: Grinding and Machining
With CBN and Diamond**

Send all inquiries to:
Glencoe/McGraw-Hill
936 Eastwind Drive
Westerville, Ohio 43081

2 3 4 5 6 7 8 9 10 11 12 13 14 15 — 00 99 98 97 96 95 94 93 92 91 90

ISBN 0-07-035587-8

The world is a user of tools. In this world of rapidly changing technology, new metal-removal tools are constantly being developed. The manufacturers who recognize the tools of tomorrow and learn how to use them today assure themselves of a place in tomorrow's prosperity. The superabrasive Borazon (CBN), one of the most promising of tomorrow's cutting tools, increases productivity, produces a better product, and lowers total grinding costs.

CONTENTS

PREFACE

Primitive people had only their hands to use as tools in order to satisfy their physical needs. However, the development of primitive tools such as stone spears and axes made their task easier. Throughout history as new tools and machines were developed, it became easier and faster to satisfy human needs. Important developments such as the machines of the industrial revolution over two hundred years ago, and more recently the computer and the second industrial revolution, have made it possible to produce goods faster and with less effort, thereby helping to raise the standard of living of most people on earth.

The successful result of scientific research is often the means by which mass production industries can effect great advances in productivity. This book is one result of the successful synthesis of diamond and cubic boron nitride (CBN) superabrasive materials now used for a wide range of industrial productivity needs. While General Electric's successful invention in 1957 of a process to manufacture diamond has not yet had the impact on metalworking manufacturing which the integrated chip and computer technology have had, the time is rapidly approaching when the use of superabrasives will be as ordinary as techniques such as computer numerical control (CNC), computer-assisted design (CAD), and computer-integrated manufacturing (CIM).

In order for this to happen, a vital role must be played by technical educators. While almost every technical school which prepares students for future manufacturing productivity has CAD laboratories, computer courses, CNC programming courses, and other facilities, only a very few are teaching the uses and applications of superabrasives in the metalworking industries. A serious knowledge gap exists between the actual capabilities of superabrasives and industries' practical abilities to apply them. This book represents a first step in closing the gap.

Cubic boron nitrate abrasives and polycrystalline diamond and CBN cutting-tool blanks have been quietly revolutionizing the grinding and machining of hard alloyed steels, metal-matrix composites, and other tough and abrasive materials used in the automotive, aerospace, and allied industries. This is only the beginning, however, as the capabilities of these super-hard, strong, tough, abrasion-resistant materials are limited only by the power and speed of the machine tools which are now available in the metalworking industries. There are already many machine tools specially designed to take full advantage of the productivity that superabrasive tools offer. Such machines are driving CBN grinding wheels at speeds over 25,000 surface feet per minute and slotting and milling tools over 10,000 feet per minute. Not only can stock be removed at previously unheard of rates, such rates are possible with the production of parts of high-precision surface finishes and surface integrity beyond that of conventional machines.

This book takes an in-depth look at the two superabrasives, diamond and CBN, and shows why it is imperative that educators and industrialists be aware of the manufacturing advantages they offer. The book is designed to acquaint the student with the scope of superabrasives which are available for metalworking, how they are made, and why their properties make them "super." Of more importance, however, is the instruction on the scope of applications related to the properties and characteristics of the workpiece materials and the type of grinding or machining process applicable. This book will enable the successful student to remove a brand-new CBN grinding wheel from its box or container and to go through all the steps of mounting, truing, and dressing the wheel for its successful use. It also covers the preparation and use of polycrystalline diamond and polycrystalline CBN tools for turning, milling, slotting, grooving, and threading operations.

To make this book as comprehensive as possible, it has been presented in 11 chapters, beginning with the development and manufacture of each superabrasive, and progresses through logical stages to the applications of each in the metalworking industry. Some of the highlights are as follows:

1. *Economics of superabrasive tools* covers the factors associated with arriving at the true cost of a manufactured part. Comparison costs are shown between conventional and superabrasive cutting tools to see which is more *cost-effective*.
2. *General machining guidelines* indicate how to set up and use superabrasive tools and what machine-tool characteristics are required for the most efficient machining operation.
3. *CBN and diamond polycrystalline* cutting tools used for turning and milling operations are covered in detail.

The purpose of this text is to provide a teaching and learning tool which is suitable for use in technical and secondary schools, colleges and universities, apprenticeship training, and industrial upgrading programs.

ACKNOWLEDGMENTS

The preparation of this book is the result of superabrasive technology "know-how" accumulated over the thirty years since these products were introduced in 1957. It would be impossible to list each person who has made a contribution in one form or another to this store of knowledge. Certainly the research team at General Electric's Corporate Research and Development Center must be accounted for in such credits. Of even more importance are the many people in the metalworking industries, in the United States and throughout the world, who had the foresight to experiment with and prove that these superabrasive tools can increase productivity and produce better-quality products.

Deep appreciation must go to the engineering and technical staff of General Electric Company Superabrasives for reviewing sections of this book and offering advice and suggestions for this project. Without the help of John Birle, Hal Bovenskerk, Tom Broskea, Alan Carius, Lynn Carrison, Tom Clark, Mark Deming, Paul Gigl, Roger Matarrese, Bob Pung, and Bill Ruark, who gave freely of their time and knowledge, this book would not have been as accurate or complete as it is. A special thanks is due to Hans Fischer of the Sunnen Products Company, who reviewed the honing section; and to Mario Rapisarda of Norwalk, Connecticut, who spent countless hours researching various sections of this book, taking special photographs, and offering many creative suggestions. Their contribution is greatly appreciated.

The authors wish to express their sincere thanks and appreciation to Alice H. Krar for her untiring devotion in reading, typing, and checking the manuscript for this text. Without her supreme effort and encouragement, this text would not have been a reality.

The authors are grateful to the following firms, which assisted with this project by supplying illustrations and technical information or by reviewing various sections of this book for accuracy and completeness.

AVCO Bay State Abrasives
Carboloy, Inc.
Carborundum Company
Cincinnati Milacron, Inc.
Exxair Corporation
GE Superabrasives
Gehring L. P. Company
Grinding Wheel Institute
Ingersoll Engineers
Landis Tool Company
Linde Division—Union Carbide.
Manufacturing Engineering
Metcut Research Associates
Norton Company
Sunnen Products Company
Vortec Corporation
Wickman Corporation
Wilkie Brothers Foundation

CHAPTER ONE

Evolution of Cutting Tools

Primitive people were always mentally superior to other animals, but nature and other animals put them at a physical disadvantage. Their keys to survival were a relatively large brain and a grasping thumb which enabled them to overcome physical handicaps and change their environment. The brain enabled them to reason and learn how to use primitive tools to improve their meager existence, and the grasping thumb enabled them to hold these tools.

For hundreds of thousands of years, tools were made from material found in nature such as stone, bone, and wood which were shaped to perform different tasks. The development of tools was a slow process. Stone axes were held in the hand for thousands of years before someone discovered that their use could be greatly improved by putting handles on them. As new tools were developed, it became easier for our ancestors to produce more food and build better shelters, but progress was painfully slow. It has been said that "Tools are the creator of civilization." If tools were not invented and developed, all of us would still be living a life of drudgery, scratching out a meager existence with basic hand tools and muscle power as our ancestors did for many centuries. It is difficult to comprehend that most of the really important breakthroughs in tool improvement have come in the last 200 years.

OBJECTIVES

After completing this chapter you should be familiar with:

1. The development of tools throughout history
2. The effects of the Industrial Revolution
3. The Law of Production
4. The development of modern cutting tools

PRIMITIVE EXISTENCE

The history of civilization is synonomous with the history of tools. Archaeologists have found that the start of community life did not begin until the introduction of tools. Early tools were made from materials found in nature such as stone, shell, tooth, horn core, bone, and wood. While some bones of animals were used as a striking weapon or ripping tool (Fig. 1-1), others were used as scoops or spoons (Fig. 1-2).

The primitives made hunting tools from small stones called "pebble tools." Pebble tools were struck together to chip away a sharp edge which would produce a crude chipping tool. Another early tool, called the *eolith* or "dawn stone," referred to as a "fist" or

"hand ax," had two edges which came together to a point to form a cutting edge (Fig. 1-3). This cutting edge enabled the user to cut branches, sharpen digging sticks, kill animals, and cut hides and edible meat for survival (Fig. 1-4). Our early ancestors lived a life of drudgery and struggled to find food and construct meager shelters with crude stone and bone tools.

There were several characteristics which set prehistoric people apart from other animals. They had a brain for the thought process, vocal cords for exchanging ideas, and hands. They soon realized that tools were an extension of the human hand, and their use made it possible to improve their standard of living.

Fig. 1-1 The lower jaw was used as a striking weapon or ripping tool. (*Courtesy of Wilkie Brothers Foundation*)

Fig. 1-3 This is called a "fist axe"—it has two edges converging to a point. (*Courtesy of Wilkie Brothers Foundation*)

Stone and Hand Tools

The simple tools that were developed during the Stone Age are grouped into three periods: the Paleolithic period or Old Stone Age, the Mesolithic period or Middle Stone Age, and the Neolithic or New Stone Age.

During the Paleolithic period or Old Stone Age, special tools were developed for the process of chipping and cutting. The flake tools, consisting of multicutting sharp edges (flakes) that were chipped from stones, were stone blades which were used as knives and spearpoints. These may have been the first cutting tools developed by prehistoric people. Smaller versions of stone blades, called *microliths*, were used as points on wooden arrows and spears. Microliths were also used as cutting edges on wooden sickles and were widely used around the beginning of the Mesolithic period or Middle Stone Age. About 25,000 years ago, primitive people discovered that by tying handles to their primitive tools they could multiply their muscle power and increase the leverage or striking force they could exert. The bow and arrow and the spear thrower (Fig. 1-5), also discovered during this period, enabled a human to throw a projectile (arrow, spear, etc.) farther and with greater force and accuracy than had previously been possible with human muscle power.

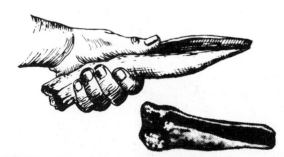

Fig. 1-2 The antelope cannon bone was used as a scoop or spoon. (*Courtesy of Wilkie Brothers Foundation*)

Farming developed during the Neolithic period or New Stone Age. Early farm tools included a wooden sickle with flint cutting edge to cut grain and a grinding stone to grind grain into flour. A stone-headed celt, which was an axlike implement, was probably used as a hoe. Plowing was greatly increased with the invention of the wooden plow, which was pulled by an animal. The invention was to coincide with the wheel, also about 3000 B.C. The Stone Age came to an end with the introduction of a hard metal—bronze; however, the standard of living due to the increased productivity each new tool provided, continued to improve.

Metal Hand Tools

The Stone Age gave way to the Bronze Age around 8000 B.C. with the discovery of copper. Copper, which is thought to have been discovered in the valleys of the Tigris and Euphrates rivers (modern Iran), became one of civilization's enduring metals. Here was a metal which could be flattened, hammered, and shaped into weapons and tools such as knives, spearheads, utensils, and agricultural tools. Once it was learned that copper could be produced in quantity by freeing the metal from the ore by heat (smelting), its fame spread. Still later, the art of casting was developed and the use of copper for making weapons, utensils, and tools for agriculture as well as for implements for clearing land became a necessity. Copper nails were used in the construction of boats and homes.

The Egyptians had made enough progress in metallurgy to alloy the malleable metal. By mixing copper and tin, they produced bronze, which was harder and more durable than copper.

About 5500 years ago, better hand tools were made out of the newly discovered copper and bronze (an alloy of copper and tin) (Fig. 1-6). With these new metal tools, our ancestors were able to produce more food and provide better shelter than had been possible before. Permanent buildings began to appear,

Fig. 1-4   Early tools used by primitive people were made of stone or wood and used to provide food and shelter. (*Courtesy of Wilkie Brothers Foundation*)

Fig. 1-5   Handles on stone tools increased our ancestors' muscle power and gave them more leverage. (*Courtesy of Wilkie Brothers Foundation*)

Fig. 1-6   Metal hand tools spurred the development of better tools and some simple mechanical devices to improve the standard of living. (*Courtesy of Wilkie Brothers Foundation*)

wheeled vehicles were developed for travel, looms made cloth weaving easier, and sailboats were built (Fig. 1-6).

The use of bronze rapidly developed during the expansion of the Roman Empire. Supplied by a workforce of slave labor and a steady shipment of copper and bronze, Rome became an important manufacturing center. Some of the treasures of this new metal included the first surgical instruments, razors, tongs, and even artificial limbs. The birth of the Iron Age began in Greece sometime between 1500 and 1000 B.C. Caesar's armies were responsible for spreading the knowledge of iron-making throughout Europe.

While many tools were cut from rock during the Stone Age and cast during the Bronze Age, tools were hammered and curved into shape during the Iron Age. Cutting edges were improved, such as the set tooth on the saw. Sickle blades and a crosscut saw, Egyptian artifacts, were uncovered during this period. In addition, the plane and shears were introduced, where the cutting action of a tool was controlled by physical means. Plowshares made of iron were developed during the early stages of the Iron Age.

Until about 1800, everything that humans ate, wore, or used was still being made with hand tools, and the basic design of these tools had not changed for thousands of years.

Machine Tools

Primitive humans' arm and hand movements set the pattern for today's machine-tool operations. The following basic machine-tool motions are directly related from work movements performed by humans since the dawn of history.

1. *Drilling* (Fig. 1-7A) consisted of a twisting motion developed from the early bow drill to the hand brace.
2. *Turning* (Fig. 1-7B) originated as the early turntable or potter's wheel used for making pottery.
3. *Planing* and *sawing* (Fig. 1-7D and 1-7E) consisted of straight-line motions with some form of hand-held cutting tool.
4. *Forming* and *forging* (Fig. 1-7G) is derived from the primitive pounding movements and refined over the years as new hand tools became available.

The most significant difference in substituting machine tools for human hands is how the workpiece and cutting tool are held and where the power to drive the cutting tool comes from (Fig. 1-8A and 1-8B). When muscle-powered hand tools are used, they produce very little very slowly, and the parts manufactured are not interchangeable. Machine tools can duplicate hand motions with far greater efficiency and speed. The machine tool holds the workpiece,

Drilling	Turning	Milling	Planing	Sawing	Grinding	Forming
(A)	(B)	(C)	(D)	(E)	(F)	(G)

Fig. 1-7 Basic machine-tool motions are derived from work motions performed by humans. (*Courtesy of Wilkie Brothers Foundation*)

which provides better accuracy, and holds and drives the cutting tool, which replaces the human muscle.

The advantages of machine tools over hand tools made marked changes in manufacturing methods. Although humans still control them, these new tools have the ability to cut, drill, pound, and reshape materials in ways that were very difficult, if not impossible, with hand tools. Machine tools could produce work faster and more accurately than could hand tools, and they were not limited by human error or weakness.

(A)   (B)

Fig. 1-8 The hand brace for drilling developed into the drill press. (A) Hand brace. (B) Radial drill press. (*Courtesy of Wilkie Brothers Foundation*)

THE INDUSTRIAL REVOLUTION

During the 1700s and early 1800s, great changes took place in the lives and work methods of people in various parts of the world. The Industrial Revolution began in Great Britain, believed to be in the late 1700s, and quickly spread throughout the world. The introduction of power-driven machinery created an enormous increase in the production of many goods. The Industrial Revolution replaced hand operations with machine operations and was responsible for taking manufacturing out of the home workshop and into factories.

Nothing has contributed more to our present standard of living and way of life than the invention of the steam engine and the boring mill, which made the invention of the modern engine possible (Fig. 1-9A and 1-9B). They gave tremendous impetus to the Industrial Revolution, which in only 175 years generated

(A)

(B)

Fig. 1-9 The invention of the boring mill made it possible to manufacture the steam engine, which gave a tremendous push toward the Industrial Revolution. (A) Boring mill. (B) Steam Engine. (*Courtesy of Wilkie Brothers Foundation*)

Evolution of Cutting Tools 5

perhaps far greater gains in the material welfare of humans than did any other invention in history.

Among the early mechanical developments in America which took the drudgery out of handwork and improved the standard of living were:

1787—Oliver Evans developed a process whereby flour was milled mechanically.

1790—Samuel Slater introduced water-powered spinning mills.

1793—Eli Whitney invented the cotton gin, which eliminated hand labor and increased the production of cotton 50 times.

1831—Cyrus McCormick developed the reaper, which could cut 600 percent more grain than a farm hand could cut with a scythe.

Machine-Tool Development

A *machine tool* may be defined as a power-driven tool used to shape or form metal by cutting, grinding, impact, pressure, electrical techniques, or a combination of these processes. *Machine tools* replaced hand tools, and *mechanical energy* replaced muscular energy. These two factors multiplied our human energy, increased productivity a millionfold, and made better products faster.

Let us take a brief look at the machine tools that made the Industrial Revolution possible.

1. *The boring mill,* invented by John Wilkinson in 1775, made it possible to drill a more precise hole. This made it possible for James Watt to manufacture a leakproof steam engine that provided a cheap, efficient source of power for the Industrial Revolution.
2. *The hacksawing machine,* developed in 1788 by Joseph Bramah, was used for slotting metal lock barrels.
3. *The screw cutting lathe,* invented by Henry Maudlay in 1800, contained a lead screw, a change of gears, and a compound rest slide. It revolutionized metalwork turning because of its versatility and marked the beginning of accurate machining.
4. *The metal planer,* developed by Richard Roberts in 1817, produced duplicate surfaces that were flat for the first time. On this machine tool, the workpiece was made to move past a stationary cutting tool.
5. *The milling machine,* invented in 1818 by Eli Whitney, used a revolving multitoothed milling cutter to machine a workpiece which was passed beneath it.
6. *The drill press,* developed by James Nasmyth in 1846, was the first power feed drill press and made the drilling of accurate holes possible.
7. *The surface grinder,* origin unknown, appeared in 1873. A Norton bonded wheel and a worktable

transferred surface finishing from hand to mechanical control.

Over the years, machine tools have been refined and new machine tools have been developed to relieve humans from tiring work and to produce more and better goods. Modern machine tools are able to produce a large quantity of parts quickly and with amazing accuracy. Computers and numerical control are now used to control the operation of machine tools and eliminate human error. Modern tools are an extension of the human hand and brain and the source of vastly improved income and material welfare.

HISTORY OF ABRASIVE TOOLS

The use of abrasive tools dates back almost two million years. Prehistoric artifacts have been discovered which indicate that our ancestors depended on tools produced by a chipping-abrading (grinding) process for their survival. It was then that humans learned that when two materials are brought together by impacting or rubbing, the softer of the two will give, yield, or chip. From this process abrasive cutting evolved.

While abrasive cutting never achieved any sophisticated level of application, it is known that the Egyptians relied heavily on natural abrasives to cut, polish, and finish the stones that were used to build the great pyramids that marked their noble civilization. There is also evidence that the Egyptians made their own grinding wheels from natural sandstone quarries and turned them on a simplified form of cylindrical grinder.

Until the latter part of the nineteenth century, abrasive tools depended on abrasives formed by nature, despite the fact that the first bonded grinding wheel (using natural abrasives) was made at the beginning of the same century. A process whereby abrasive grains could be mass-produced in a highly controlled environment was needed. This came about with the introduction of the electric furnace process for producing silicon carbide. The development of manufactured aluminum oxide grain came a short time later.

Today, aluminum oxide and silicon carbide abrasive grains can be perfectly matched by size and toughness property to suit a variety of grinding situations and operating conditions. The grinding machines themselves may be computer numerically controlled and have automatic feeding, dressing, and wheel-wear compensation features. Some of the most advanced systems feature robotically operated grinding wheel changes.

Many of the materials that industry must work with have required abrasive grains which are harder and stronger than aluminum oxide and silicon carbide. While natural-diamond abrasives have been used over the past 100 years to manufacture saw blades used in the natural-stone industry, diamond abrasives have been used extensively in the manufacture of grinding wheels for grinding cemented carbides, ceramics, and glass only in the last 50 years or so. The use of diamond as an abrasive clearly illustrates the technical and productivity advantage which an abrasive of extraordinary hardness and strength has in these processes.

By 1955, General Electric Company announced that it had successfully manufactured diamond abrasive and appropriately used the trademark *Man-Made* diamond* to identify this product. This was to be the first in a long series of products for which the company had coined the term *superabrasives*.

Along with the search to find a method of synthesizing diamond came the synthesis of a unique material—cubic boron nitride. Cubic boron nitride has a hardness which is superior to any other known material except diamond itself. With its hardness and strength characteristics, it offers a significant potential as a new abrasive product.

Even though grinding is the oldest manufacturing process known to civilization, it remains a fundamentally sound modern process that continues to grow in its scope of capabilities.

Metal-Cutting-Tool Development

A large percentage of metal machined is cut on a lathe or a milling machine. Since the variety of operations performed on these two machines varies greatly, an assortment of cutting tools and shapes are used for machining. The two general types of cutting tools used are single-point tools, used primarily for turning (lathe) operations, and multipoint tools, used primarily on milling machines.

One of the most important factors in the machining process is the cutting tool; its performance will determine the efficiency of the operation. The cutting tool can make or break an operation or an entire manufacturing process. The best cutting tool should always be used which provides a balance between cutting-tool life and the maximum production rate. *The cost of a cutting tool is insignificant; the effect of the cutting tool can be enormous.*

As new metals were discovered throughout history, new cutting tools were developed to cut them more efficiently. A brief history of metal cutting tools is shown in Fig. 1-10. As can be seen, new cutting materials were developed to meet the machinery requirements for new metals and to improve productivity.

*Trademark of General Electric Company, USA.

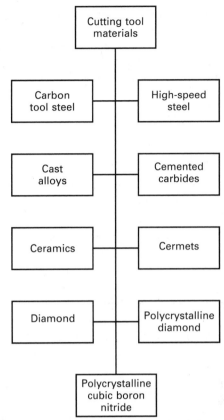

Fig. 1-10 The development of metal cutting-tool materials.

1. *Carbon tool steels*, made of high-carbon steel, were used for machining metals until they were replaced by high-speed steel. They are not used today for machining metal for industrial purposes because the cutting edge breaks down very quickly.
2. *High-speed steels* may contain varying amounts of tungsten, chromium, vanadium, molybdenum, and cobalt. They can withstand more than three times the heat of high-carbon steel tools, take heavy cuts, withstand shock, and maintain a sharp cutting edge at high cutting speeds.
3. *Cast alloys* contain chromium, tungsten, carbon, and cobalt. They have high hardness, high resistance to wear, and excellent hot-hardness qualities. They can be operated at speeds two to two and a half times faster than high-speed steel tools.
4. *Cemented carbides* contain tungsten carbide and cobalt, with varying amounts of titanium or tantalum carbide. They have low toughness, lower than high-speed steel, but high hardness and excellent hot-hardness qualities. They can be operated at cutting speeds three to four times higher than high-speed steel tools. *Coated carbide tools* have a thin layer of titanium carbide or aluminum oxide fused to the cutting tool. This wear-resistant coat-

ing provides for longer tool life and permits the use of higher cutting speeds.

5. *Ceramics* consist of mainly aluminum oxide blended with additions of titanium oxide or titanium carbide. Ceramic cutting tools permit higher cutting speeds, increased tool life, and better surface finish than do carbide tools.

6. *Single-crystal natural-diamond* tools are used mainly to machine nonferrous metals and abrasive nonmetallic materials. They can be operated at very high speeds and produce excellent surface finishes. Diamond-cutting tools have excellent wear qualities but have a very low shock-resistant factor.

7. *Polycrystalline diamond* (PCD) consists of very fine manufactured diamond crystals sintered and

bonded on a carbide substrate. Tools made of PCD are used for machining nonferrous metals and abrasive nonmetallic materials at speeds much higher than cemented carbide tools. They have excellent wear and shock resistance and may provide up to 100 times greater tool life than carbide tools.

8. *Polycrystalline cubic boron nitride* (PCBN) consists of a layer of very fine cubic boron nitride crystals sintered and bonded to a carbide substrate. PCBN tools are used for machining high-temperature alloys and ferrous alloys. They have exceptional wear resistance and edge life and good shock-resistance qualities. Materials as hard as Rockwell C68 can be machined quickly, eliminating the need for the slow operation of grinding.

Everything contributing to human material welfare is the product of natural resources plus human energy multiplied by tools.

NR + HE X T = HMW

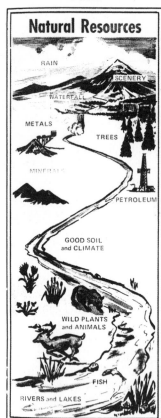

Natural Resources

RAIN
SCENERY
WATERFALL
METALS
TREES
MINERALS
PETROLEUM
GOOD SOIL and CLIMATE
WILD PLANTS and ANIMALS
FISH
RIVERS and LAKES

Human Energy

Tools

MACHINE TOOLS • FARMS MILLS and FACTORIES

ELECTRICAL and MECHANICAL ENERGY AND POWER TRANSMISSION

TRANSPORTATION and COMMUNICATION

MONEY BANKING CREDIT INSURANCE
DEPARTMENT STORE

EXCHANGE AND MARKET FACILITIES

Human Material Welfare

FOOD   CLOTHING   HOUSING

EDUCATION
ENTERTAINMENT
PUBLIC HEALTH
TRANSPORTATION
LABOR-SAVING DEVICES
COMMUNICATION
SERVICES
LAUNDRY
GOVERNMENT SERVICES   LAW and ORDER   NATIONAL SECURITY
POLICE

Natural Resources include all of the things made available by nature. These are useless until man has changed their form, condition, or place.

Human Energy, both mental and muscular, even when aided by animal power, are not enough to produce any more than a bare living.

Modern Tools must be thought of as including everything used by man to produce and exchange goods and services. Even money and credit are tools.

Human Material Welfare, in addition to food, clothing, and shelter, includes all of the other benefits made possible by an economic surplus.

Fig. 1-11 The Law of Production consists of natural resources, human energy, and tools to determine human material welfare. (*Courtesy of Wilkie Brothers Foundation*)

LAW OF PRODUCTION

To understand how a nation can produce more goods and services each year at lower cost, we must examine the term *output per labor-hour*. In industry, every operation is continuously examined to find better ways to improve the output of products or services and lower production costs. When better machine or cutting tools are developed, manufacturers must take advantage of these more productive tools or risk losing their customers to competitors who use the new technology and are able to produce the same goods for less cost.

Economic growth depends entirely on increasing the potential output per labor-hour. In a free economy, the elements of profits, wages, prices, and productivity keep in balance if not thrown out of balance by government decree or demands from a pressure group. Productivity is the foundation of a sound economy. To overspend beyond its ability to produce is a form of self-destruction that can bleed a nation dry monetarily. One need not look further than to examine the British economy in the 25-year period before World War I. Their productivity fell behind the Americans and Germans, who surpassed them technologically.

The Law of Production is best summed up in Fig. 1-11, which states: *Everything contributing to a human's material welfare is the product of natural resources plus human energy multiplied by tools.* The equation reads:

$$NR + HE \times T = HMW$$

where
NR = natural resources
HE = human energy
T = tools
HMW = human material welfare

TOOLS FOR PRODUCTIVITY

In our modern economy, it is important to recognize that everything used in production and exchange is a tool. The purpose of a factory is to house the tools, but it is the land, buildings, and all nonproductive equipment that make the use of power tools possible. Everything owned by a company must be considered a tool which is necessary to produce products.

These tools have generally been purchased with the money supplied by stockholders who invested their savings in the company. We must all understand the profit system because profit is the amount collected from the customer on behalf of the stockholders for the use of the tools—meaning all the assets. Therefore, profit must be thought of as payment for the use of the tools. Stockholders would not invest in a company if there were to be no payment for the use of their savings and, as a result, there would have been no tools and no business. Therefore, profit is the most important of *business costs* and the most vital part of the selling price. Although tools made the United States great, it is the acceptance of the profit system that caused the tools to be created.

The prosperity of any country depends not only on better tools but also the freedom with which they are used. Tools are responsible for our affluent standard of living directly through their productivity. We continually develop new tools, yet in the past two decades we have been reluctant to use them until someone else has proven their worth. Countries such as Japan and Germany have implemented the tools developed in the United States and have learned how to produce better-quality products, at a lower price, to capture an increasingly greater share of our and world markets. The United States has lagged behind in making use of the technology Americans created. This trend must be reversed before it is too late.

REVIEW QUESTIONS

1. Name three materials which were used by primitive people as tools.

Primitive Existence

2. What was the eolith or dawn stone?
3. Name the three characteristics which set humans apart from animals.

Stone and Hand Tools

4. List the most important developments in tools during the three Stone Age periods.

Metal Hand Tools

5. Why was the development of copper so important to the manufacture of tools?
6. What advantage did bronze have on hand tool manufacture?

Machine Tools

7. List four advantages which machine tools have over hand tools.

The Industrial Revolution

8. Name the two inventions which gave tremendous impetus to the Industrial Revolution.

9. List the four early mechanical developments in early America which improved the standard of living.

Machine-Tool Development

10. What effect did machine tools have on the process of manufacturing parts?

History of Abrasive Tools

11. What two developments in the latter part of the nineteenth century led to great advances in grinding and grinding machines?
12. Briefly trace the development of manufactured diamond and cubic boron nitride.

Metal-Cutting-Tool Development

13. What is one of the most important factors in the machining process? Explain why it is important.
14. List four of the most important cutting-tool materials.

Law of Production

15. What does economic growth depend on?

Tools for Productivity

16. Name the things that are considered tools in our modern economy.

CHAPTER TWO

Manufacture of Superabrasives

In the continuing efforts to improve product quality and also manufacturing methods, new machine tools and new materials have been developed. Some of these new metals (superalloys) and composite materials have proved to be very difficult, if not impossible, to grind or machine with conventional tools. The manufacture of new abrasives (superabrasives) such as Man-Made* diamond and CBN (cubic boron nitride) has made it possible to effectively grind and machine difficult-to-cut steels, superalloys, and composite materials while still holding tolerances (part size) and geometry (shape).

As exciting as these new high-efficiency cutting tools promise to be, they cannot be applied to every grinding or machining application. However, they have proved to be exceptional in the machining of hard ferrous metals, cast irons, and nickel-base and cobalt-base superalloys. These new high-efficiency superabrasive cutting tools can produce more goods, to a higher degree of accuracy for the same physical effort and at a lower cost, which, in turn, increases productivity.

OBJECTIVES

After completing this chapter you should be able to:

1. Describe the manufacture of industrial diamonds
2. Identify the properties of manufactured diamond and how it compares to other abrasives
3. Describe the manufacture of cubic boron nitride (CBN)
4. Identify the properties of cubic boron nitride (CBN) and its use in the metalworking industry

PROPERTIES OF SUPERABRASIVES

The superabrasives, diamond and CBN, possess properties unmatched by conventional abrasives such as aluminum oxide and silicon carbide. The hardness, abrasion resistance, compressive strength, and thermal conductivity of superabrasives make them logical choices for many difficult grinding, sawing, lapping, machining, drilling, wheel-dressing, and wire-drawing applications. Superabrasives can penetrate the hardest materials known to humans, making difficult material-removal applications routine operations. Table 2-1A compares three important properties of the superabrasives with those of the two most common conventional abrasives, aluminum oxide and silicon carbide. Cemented tungsten carbide, an important workpiece for diamond grinding, is included for comparison.

*Trademark of General Electric Company, USA.

Hardness Measurement

Hardness is defined as resistance to local penetration. "Resistance to local penetration" will depend on several properties of the material such as yield strength, elasticity, and work hardening characteristics. For most industrial purposes, hardness is measured with an indentation hardness method.

This text uses four different hardness scales to describe the properties of superabrasives and the workpieces on which they are used.

There are three common systems for making indentation hardness tests:

Brinell Hardness Test. A hard ball is forced into the surface to be tested. The diameter or depth of the impression is measured, and the *Brinell hardness number* (BHN) is calculated from this measurement. This test cannot be used on very thin materi-

Table 2-1A MECHANICAL PROPERTIES OF HARD MATERIALS

Material	Knoop Hardness	Compressive Strength, psi	Thermal Conductivity, cal/(°C · cm · s)
Diamond	8000	1,260,000	5.0
Cubic boron nitride	4500	945,000*	3.3
Aluminum oxide	2500	500,000	0.08
Silicon carbide	2700	210,000	0.10
Tungsten carbide	2100	500,000	0.08

*Estimate

als, very hard materials, or materials with a significant change in hardness below the material surface.

Rockwell Hardness Test. Five different combinations of applied load and penetrator may be used in the Rockwell system. The penetrator is either a small hard ball or a special diamond point. Each of the five combinations of load and penetrator is designed to accurately measure hardness for materials of a specific hardness range. Thin, small samples can be measured using the Rockwell system. In this text, only two of these ranges or scales are used:

- **Rockwell "A" Scale.** This scale is used only for very hard metals. A very small diamond cone is used as the penetrator. Metals measured on this scale may be identified as Ra 88 (Rockwell "A" scale hardness of 88).
- **Rockwell "C" Scale.** The same diamond penetrator is used as for the "A" scale but with a different load. This scale is used for almost all types of steels, including those which have been hardened. Metals measured on this scale may be identified as Rc 60 (Rockwell "C" scale hardness of 60).

Knoop Hardness Test. This test is generally for laboratory use and is not suited for normal industrial hardness testing. This test uses a very special pyramidal diamond penetrator, and with very careful use, can actually measure the hardness of diamond itself. Its scale is identified as *Knoop Hardness Number* (KHN) and has the units of kilograms per square millimeter (kg/mm^2).

A reference conversion chart of these four scales is shown in Table 2-1B.

The main physical properties of superabrasives are illustrated in Fig. 2-1A.

Hardness. The hardness property is very important for an abrasive. The harder the abrasive with respect to the workpiece, the more easily it can cut.

Table 2-1B HARDNESS CONVERSION CHART

Rockwell Hardness Scales		Knoop Hardness Number	Brinell Hardness Number
C	A	KHN	BHN
80	92.0	1865	—
79	91.5	1787	—
78	91.0	1710	—
77	90.5	1633	—
76	90.0	1556	—
75	89.5	1478	—
74	89.0	1400	—
73	88.5	1323	—
72	88.0	1245	—
71	87.0	1160	—
70	86.5	1076	—
69	86.0	1004	—
68	85.5	942	—
67	85.0	894	—
66	84.5	854	—
65	84.0	820	—
64	83.5	789	—
63	83.0	763	—
62	82.5	739	—
61	81.5	716	—
60	81.0	695	614
59	80.5	675	600
58	80.0	655	587
57	79.5	636	573
56	79.0	617	560
55	78.5	598	547
54	78.0	580	534
53	77.5	562	522
52	77.0	545	509
51	76.5	528	496
50	76.0	513	484
49	75.5	498	472
48	74.5	485	460
47	74.0	471	448
46	73.5	458	437
45	73.0	446	426
44	72.5	435	415
43	72.0	424	404
42	71.5	413	393
41	71.0	403	382
40	70.5	393	372
39	70.0	383	362
38	69.5	373	352

Table 2-1B (continued)

Rockwell Hardness Scales		Knoop Hardness Number	Brinell Hardness Number
C	A	KHN	BHN
37	69.0	363	342
36	68.5	353	332
35	68.0	343	322
34	67.5	334	313
33	67.0	325	305
32	66.5	317	297
31	66.0	309	290
30	65.5	301	283
29	65.0	293	276
28	64.5	285	270
27	64.0	278	265
26	63.5	271	260
25	63.0	264	255
24	62.5	257	250
23	62.0	251	245
22	61.5	246	240
21	61.0	241	235
20	60.5	236	230

Figure 2-1A compares the hardness of diamond and cubic boron nitride with the same properties of the two major conventional abrasives. The Knoop hardness test is a standard method of measuring the hardness of exceptionally hard and brittle materials and individual grains and particles. It is an indentation test and thus is regarded as a true test of the relative hardness of materials. Diamond between 7000 and 10,000 on the Knoop hardness scale is the hardest known substance, while CBN at 4700 on the scale is second only to diamond in hardness.

Abrasion Resistance. Figure 2-1B shows the relative abrasion resistance of diamond and cubic boron nitride in relation to the conventional abrasives. Diamond has about three times the abrasion resistance of silicon carbide, while CBN has about four times the abrasion resistance of aluminum oxide. This high abrasion resistance makes superabrasive cutting tools ideal for machining hard, tough materials at high cutting speeds. In comparison to conventional tools, superabrasive tools maintain their sharp cutting edges much longer, thereby increasing productivity while at the same time producing parts which are dimensionally accurate.

Compressive Strength. Figure 2-1C illustrates the compressive strength of superabrasives and conventional abrasives. *Compressive strength* is defined as the maximum stress in compression that a material will take before it ruptures or breaks. The high compressive strength values are to be expected because both diamond and CBN crystals have nearly the same density since they have almost the same inner structures. Therefore, superabrasive tools have excellent qualities to withstand the forces created during high metal removal rates and the shock of severe interrupted cuts.

Thermal Conductivity. Figure 2-1D shows that diamond and CBN superabrasives have excellent thermal conductivity which is much higher than aluminum oxide or silicon carbide. This allows greater heat dissipation (transfer), especially when cutting hard, abrasive, or tough materials at high material-removal rates. The high cutting temperatures created at the cutting-tool–workpiece interface would weaken or soften conventional cutting-tool materials. Heat generated during the operation is rapidly dissipated through the superabrasive material into the grinding wheel or tool, thus reducing the risk of thermally damaging the workpiece material.

Because of the unique combination of properties—hardness, abrasion resistance, compressive strength, and thermal conductivity—the superabrasives have achieved a position of major importance in many industries. Four major factors account for the broad acceptance and continuing growth in the use of these truly "super" materials.

1. The increasing precision of modern machining operations has increased the use of superabrasive tools. A superabrasive tool keeps its cutting edge almost unchanged throughout most of its useful life, holding the tolerances set, and in many instances reducing scrap to zero or near zero.
2. Automation has placed great importance on continuity of production. As an operation becomes more automatic, machine downtime becomes more costly. It is not unusual for a properly designed superabrasive tool to produce 10 to 100 times the number of pieces formerly produced by conventional tooling.
3. The new workpiece materials, many of them very hard and/or abrasive, have created cutting problems that often can be solved only by superabrasive tools.
4. More people are learning to use superabrasive tools, and superabrasive tools are not only used for working "exotic" materials. In many applications, superabrasives are performing the same jobs formerly done by conventional abrasives or tool materials, but they are doing it better, faster, and at a lower cost.

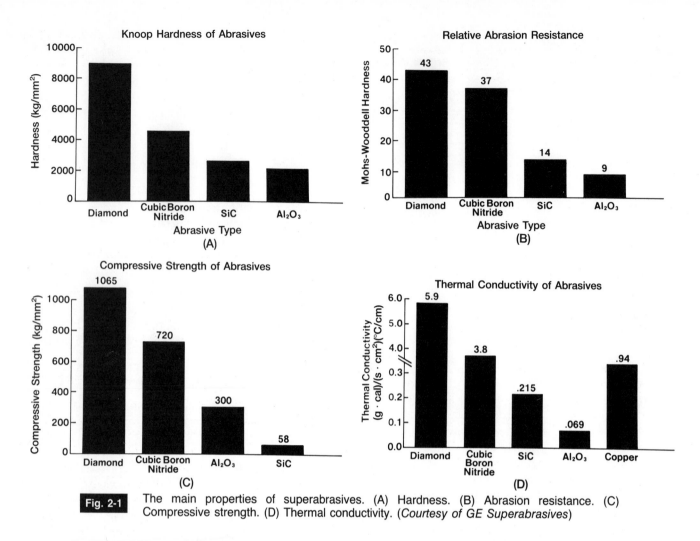

Knoop Hardness of Abrasives

Relative Abrasion Resistance

Compressive Strength of Abrasives

Thermal Conductivity of Abrasives

Fig. 2-1 The main properties of superabrasives. (A) Hardness. (B) Abrasion resistance. (C) Compressive strength. (D) Thermal conductivity. (*Courtesy of GE Superabrasives*)

CARBON, GRAPHITE, AND DIAMOND

Of all the chemical elements, carbon is the most important to humanity. Combined with other elements, it is the basis of all living beings. In this section, the element carbon itself will be examined.

Carbon occurs in three common forms: *amorphous carbon* (glassy carbon or carbon black) and two *crystalline forms,* graphite and diamond. The arrangement of carbon atoms in space with respect to each other determines the form. In amorphous carbon, there is no order to the arrangement. In graphite (Fig. 2-2A) the atoms form planes of hexagons stacked one above the other. Each atom attaches to three others in the plane. They bond very tightly within the planes and loosely between planes. For this reason, graphite is slippery to the touch and is often used as a lubricant. In diamond (Fig. 2-2B) the atoms of carbon closely pack in three dimensions. Each carbon atom attaches to four others, forming a hard, dense structure—the hardest material known.

Carbon is in the middle of the periodic table of the elements (Fig. 2-3). To its left are the metals and to its right, the nonmetals. One might infer that carbon behaves somewhere between the two. In fact, *graphite*

behaves a little like metal; it conducts electricity, for instance. *Diamond,* on the other hand, behaves more like a nonmetal; it is an electrical insulator but conducts heat very well.

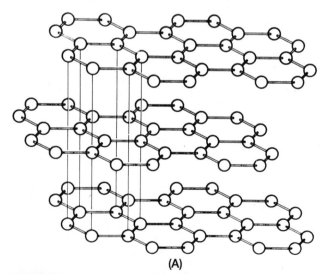

(A)

Fig. 2-2A The crystallographic structure of graphite. (*Courtesy of GE Superabrasives*)

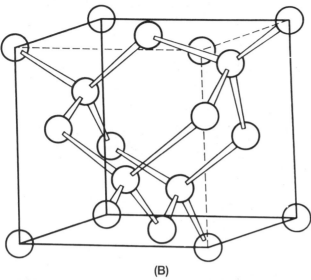

(B)

Fig. 2-2B The crystallographic structure of diamond. (*Courtesy of GE Superabrasives*)

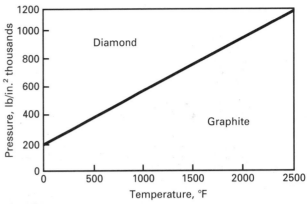

Fig. 2-4 The stability regions for diamond and graphite. (*Courtesy of GE Superabrasives*)

At atmospheric pressure, graphite is the stable crystalline form of carbon. Graphite does not melt when heated in the absence of oxygen. Above 3500°C (6330°F) it transforms from a solid to a gas; in air or oxygen, it burns to form carbon dioxide. Diamond is truly stable only at high pressures. Figure 2-4 shows the areas of temperature and pressure where graphite and diamond are stable. Once diamond forms, it remains diamond when cooled and brought down to one atmosphere. Scientists do not know exactly how diamond crystals grow in nature; however, it is certain that they form in the area of pressure and temperature where diamond is stable. This means that diamond starts very deep in the earth and rises to the surface by massive movements of material similar to a volcanic eruption.

In order to produce diamond by manufacturing processes, the conditions of pressure and temperature found far below the earth's surface must be duplicated. This in itself is a challenge requiring in-

Metals →   ← Non—Metals

B^5 C^6 N^7

Boron   Carbon   Nitrogen

Al^{13} Si^{14} P^{15}

Aluminum   Silicon   Phosphorus

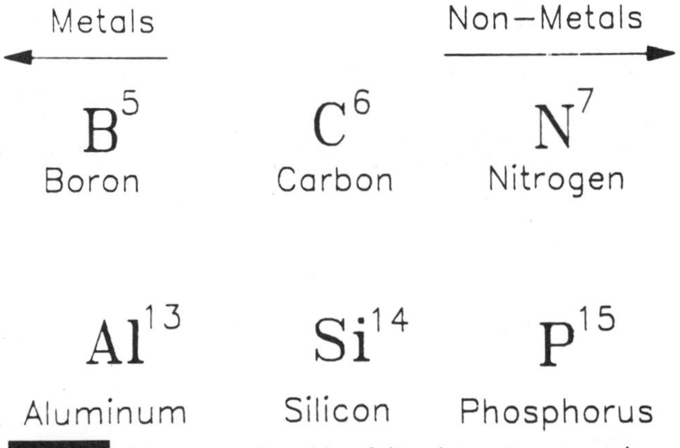

Fig. 2-3 The periodic table of the elements near carbon.

novative design of high-pressure, high-temperature apparatus. In such apparatus, graphite is transformed into diamond. The diamond is then brought back to room temperature and atmospheric pressure intact.

DEVELOPMENT OF THE DIAMOND PROCESS

As early as 1880, people were experimenting with various methods attempting to manufacture diamond. The early workers knew that making diamond required high pressures and temperatures, but not the exact conditions. The technology available to them was very crude; none were successful.

In 1951 it was clear that the United States required a domestic source of industrial diamond. In the preceding decade, cemented tungsten carbide tools became an essential part of metalworking processes. Only diamond could economically sharpen these tools, but there existed a broad range of other needs for diamond in sawing and drilling stone and concrete as well as dressing tools for grinding wheels. The United States was totally dependent on natural diamond, all of which was available only through overseas sources. For this reason, the General Electric Company (GE) committed significant resources to the development of a manufactured diamond process.

The team of scientists and engineers formed to develop the diamond process faced many difficult challenges. By far the greatest challenge was building an apparatus capable of containing 1 million pounds per square inch (lb/in.²) and 1600°C (2900°F); such pressures exist only 160 miles (mi) or more below the earth's surface. Scientists had never been able to duplicate these extreme conditions in the laboratory, but after many attempts and failures, the team designed the *belt apparatus* (Fig. 2-5), which uses cemented tungsten carbide pistons or punches and a cemented tungsten carbide die or "belt." The punches and die are prestressed to the maximum limit possible in special high-toughness steel rings pressed together with interference fits. By using the concept of prestressing,

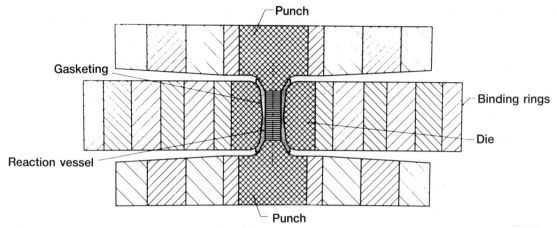

Fig. 2-5 The high-pressure high-temperature apparatus used to manufacture diamond. (*Courtesy of GE Superabrasives*)

much of the applied stress is counteracted when the vessel is in use. Without this prestressing, the die and punches would break long before the desired pressure was reached. In addition to the pressure vessel itself, the design of the gasketing system presented a challenge. It had to contain the materials used for diamond making, transmit pressure uniformly, insulate the carbide parts from the high temperature, and seal the system. A composite structure of the mineral *pyrophyllite* and steel met these criteria. By 1953, the pressure system was ready.

Having developed the pressure vessel and gasketing system, the real mission began. Many carbon-containing materials were subjected to pressures and temperatures well within the range where diamond is stable. For many experiments, no diamond formed. The key discovery was the concept of a molten metal *solvent-catalyst*. The use of molten iron, nickel, cobalt, or mixtures of these metals in combination with a pure source of carbon such as graphite produced the first diamond ever manufactured. The GE team in Schenectady, New York, reported success on February 15, 1955.

Making a few crystals of diamond in the laboratory is a long way from having a product for grinding tungsten carbide. Much development and testing took place before a commercial product became available for use. To obtain the right diamond properties for a given application, it was necessary to select the right graphite, the proper solvent-catalyst, and the right conditions of pressure, temperature, and time. GE introduced the first product, called Type A Man-Made diamond, in October 1957. With several intermediate improvements, this evolved into an elongated friable crystal with rough edges, commonly called RVG* diamond (Fig. 2-6A). The properties of RVG make it ideal for use in resin and vitreous grinding wheels for grinding tungsten carbide. Wheel

*Trademark of General Electric Company, USA.

manufacturers throughout the world continue to regard this diamond crystal as the standard of the industry for this application. Since 1957, GE has developed and introduced other diamond types.

MBG* (metal bond grinding), a tough, block-shaped diamond crystal not as friable as the RVG type, is used for glass and ceramic processing (Fig. 2-6B).

MBS* (metal bond saw) diamond, a blocky, extremely tough crystal with a smooth regular surface, is used in metal-bonded saws for cutting stone and concrete (Fig. 2-6C).

For the metalworking industry, however, RVG for tungsten carbide grinding is the product of interest.

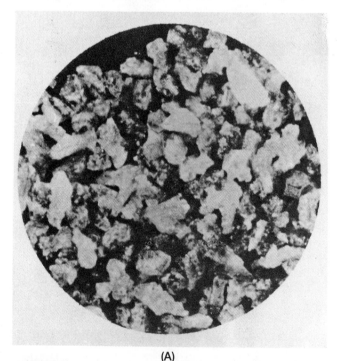

(A)

Fig. 2-6A The RVG diamond is used to grind ultra-hard materials. (*Courtesy of GE Superabrasives*)

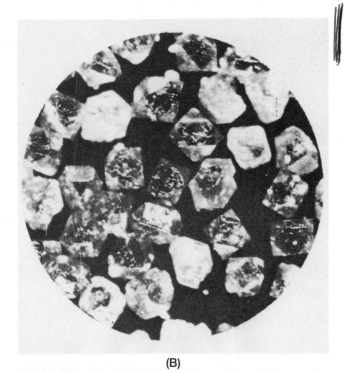

(B)

Fig. 2-6B The MBG diamond is used to grind cemented carbides, glass, ceramics, and other materials. (*Courtesy of GE Superabrasives*)

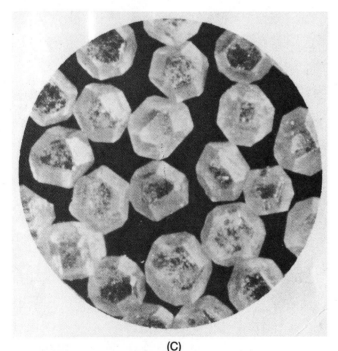

(C)

Fig. 2-6C The MBS diamond is used to cut stone, marble, concrete, and other materials. (*Courtesy of GE Superabrasives*)

The Function of the RVG Diamond Crystal

The unique property of RVG that makes it suitable for grinding tungsten carbide is *multicrystallinity*. Thousands of tightly bonded small diamond crystals make up each abrasive grain. This gives the grain a characteristic known as *friability*, the ability to break down in a controlled manner. Any diamond crystal begins to dull as it works; a very tough single diamond crystal would become dull, polish, and fail to grind rather quickly. RVG diamond crystal, because it is multicrystalline, fails by microchipping. Very small (minute) dulled pieces of diamond break off, leaving a large number of fresh cutting points on each crystal. Figure 2-7 shows such a crystal in the surface of a used grinding wheel. The wheel is sharp, open, and free-cutting. The mode of breakdown of RVG also permits grinding to a very smooth surface finish.

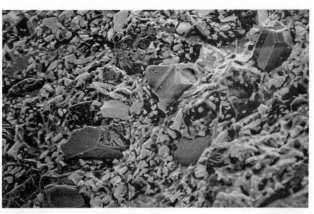

Fig. 2-7 A 140/170-grit resin bond (RVG) wheel showing the wear on the abrasive crystal. (*Courtesy of GE Superabrasives*)

Types of RVG Abrasive

In any grinding system, the abrasive itself is not the only factor controlling the cost of the grinding process. The entire system of abrasive, bond, machine, coolant, cutting rates, and workpiece are important factors. An improvement in any one part may require modifications in the rest. The RVG diamond abrasive grinds cemented tungsten carbide both wet and dry very economically. However, some wheel wear results from crystals pulled out from the resin bond; this is particularly true at high material removal rates. To provide better holding power (retention) of the RVG crystal in the resin bond, a special nickel coating was developed in 1966. This strong thick metal coating sticks (adheres) to all corners and crevices in the polycrystalline grain much better than does the resin from the bond. The specially controlled rough surfaces of the coating adheres very well in turn to the resin, thus reducing pulled-out crystals considerably. Figures 2-8 and 2-9 show the RVG-W (resin, vitrified grinding, wet) crystals and the worn surface of an RVG-W grinding wheel. High temperature at the crystal-resin interface in the presence of coolants tends to break down the resin, causing loss of crystals from the wheel. The nickel coating is much less ther-

An 80/100 abrasive grain RVG-W wheel with 56 wt % nickel coating. (*Courtesy of GE Superabrasives*)

mally conductive than diamond itself and provides a cooler surface to the resin, less deterioration, and better retention of grains. The nickel-coated RVG product is called *RVG-W*. The RVG-W crystal is available in both 56 and 30 weight percent (wt %) nickel coating, meaning that the coating makes up either 56 or 30 percent of the total weight of the grain. However, coated diamond is available on the basis of the *contained weight* of diamond. The 56 percent product is used primarily for wet grinding of cemented tungsten carbide, while the 30 percent product is used for carbide-steel composites where the steel is a significant fraction of the total tool surface being ground.

Although some improvement over RVG is noted in dry grinding with RVG-W, the mechanism is different. As there is no coolant, more heat is dissipated in the wheel. At high grinding rates, the overall temperature near the surface becomes sufficiently hot to degrade the resin, causing loss of crystals. Using a 50 wt % coating of copper on the RVG grains achieves the same holding power in the bond as does the

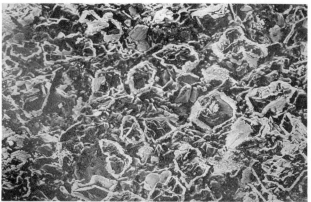

Fig. 2-9   A 140/170-resin-bond RVG-W wheel (56 wt % nickel coating) showing the wear on the abrasive crystal. (*Courtesy of GE Superabrasives*)

nickel coating. An additional benefit is an increase in the overall thermal conductivity of the wheel; this keeps the resin cooler and less likely to lose its holding qualities. The *RVG-D* crystal is designed for dry grinding with resin bonds at higher material removal rates without excessive loss of pulled-out grains. Scanning electron microscope pictures of RVG-D crystals and the worn surface of a grinding wheel containing RVG-D are shown in Figs. 2-10 and 2-11.

Development of these coated grains has allowed wheel manufacturers to develop new and improved bond systems. This further increases the productivity of the RVG grinding system.

Fig. 2-10   A 80/100 abrasive grain RVG-D crystals. (*Courtesy of GE Superabrasives*)

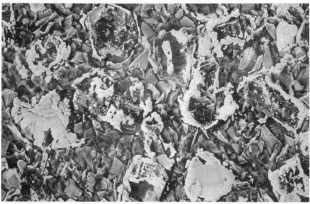

Fig. 2-11   A 140/170 resin-bond RVG-D wheel showing the wear pattern on the abrasive crystal. (*Courtesy of GE Superabrasives*)

DIAMOND AND STEEL GRINDING

One might ask why diamond is not used to grind steel. It is important to understand that diamond can be used effectively only on nonferrous (containing no iron) workpiece materials. Most steels, especially low-carbon steels, have a chemical characteristic known as *carbon solubility potential:* in other words, such steels are ready to react with any source of free

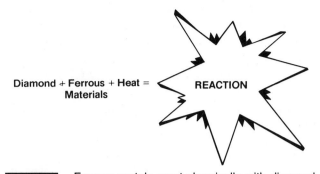

Diagram: Diamond + Ferrous Materials + Heat = REACTION

Fig. 2-12 Ferrous metals react chemically with diamond cutting tools under the temperature and pressure conditions occurring in the grinding process. (*Courtesy of GE Superabrasives*)

carbon and absorb this carbon into their surface. Such a reaction is easily triggered under the temperature and pressure conditions which normally occur in the grinding and machining process (Fig. 2-12). An analogy could be made by imagining the process of grinding frozen water (ice) with a grinding wheel which contained rock salt as the abrasive grains. In this example, the rock salt would wear primarily by the process of going into solution with the ice itself far more than through a mechanical wear process. Thus the economic feasibility of applying diamond in the grinding or machining of steel is not determined by the "super" hardness and strength of diamond, but rather by its Achilles' heel of chemical solubility in iron. This generates excessive wear of diamond and makes its use uneconomical.

DEVELOPMENT OF CUBIC BORON NITRIDE

Figure 2-3, which focuses on the periodic table of the elements near diamond (carbon), shows boron to the left and nitrogen to the right. If these were combined into a tightly bonded structure such as diamond, the crystals might be nearly as hard or even harder. The search for a new hard material led to the development of a new superabrasive not found in nature—cubic boron nitride, which was named *Borazon* * *CBN*.

Before the synthesis of cubic boron nitride (CBN) at GE's Corporate Research and Development Laboratory in 1957, a form of boron nitride (known as *hexagonal boron nitride*) much like graphite was commercially available. The arrangement of atoms is similar to graphite, shown in Fig. 2-2, with alternating boron and nitrogen instead of all carbons. This suggests that an arrangement like that of diamond with alternating boron and nitrogen should also exist. The challenge, as with synthetic diamond, was finding the right solvent-catalyst system. The GE inventors discovered that alkali metals and nitrides of lithium, calcium, and magnesium worked best. Many other nitrides or

*Trademark of General Electric Company, USA.

metals also work but are less effective. As predicted, the analog of diamond, cubic boron nitride (CBN), was produced.

Those working on the project had hoped to make something harder than diamond and at first were disappointed. The hardness is somewhat less; CBN did not grind tungsten carbide very well. Only when the need for a noncarbon superabrasive for grinding hardened steel was clear and the properties of this newly developed abrasive determined did a new product arise. GE commercially introduced Borazon CBN Type I (uncoated) and Type II (60 percent nickel-coated) in 1969.

The impact of CBN on the steel-grinding process is as great as that of using diamond to replace silicon carbide for grinding cemented tungsten carbide. On hardened tool steels, the performance of CBN is far superior to that of the conventional abrasive, aluminum oxide.

TYPES OF CUBIC BORON NITRIDE

Like diamond, CBN is specially manufactured into a variety of products for different applications. Unlike diamond, all CBN products are used in the metal-working industry. There are currently seven different types of Borazon CBN abrasives available, each manufactured to specifically suit one or both of the following major criteria: (1) bond system and (2) mode of grinding and material-removal rate.

Type I. This is a medium-toughness monocrystalline CBN abrasive designed for use in vitrified, metal, and electroplated-bond grinding wheels (Fig. 2-13A).

Type II. The surfaces of the Type I crystal are too smooth to be held securely in a resin bond; therefore, they must be coated with a heavy nickel metal coating to make them suitable for use in

Fig. 2-13 Cubic boron nitride 80/100 abrasive crystals. (A) Type I. (B) Type II. (*Courtesy of GE Superabrasives*)

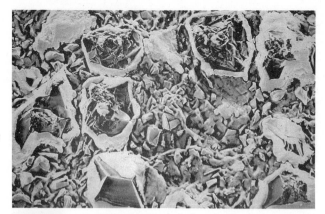

resin-bond systems. Type II is a Type I crystal which has a 60 wt % nickel coating covering its surfaces. This coating tailors the Type II abrasive solely for use in resin-bond systems. Since resin-bonded wheels are the most frequently used type of CBN grinding wheel today, Type II is the most widely used CBN abrasive (Fig. 2-13B). Both Type I and Type II crystals wear by the controlled *cleavage* of the grains leaving sharp cutting points. Figure 2-14 illustrates the surface of a CBN Type II wheel showing this wear pattern.

Type 500. This is a high-toughness monocrystalline CBN abrasive suited specifically for more demanding applications in vitrified and electroplated-bond systems (Fig. 2-15A).

Type 510. This abrasive grain is manufactured by applying a thin coating of titanium to the Type 500 crystal. The purpose of this coating is to improve the retention of the Type 500 crystal in metal and vitrified-bond systems where a stronger matrix-crystal bond is necessary for high material-removal rates (Fig. 2-15B).

**Monocrystalline
CBN Particle Wear Pattern**

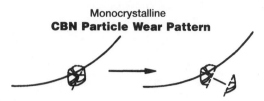

Fig. 2-16 | Monocrystalline CBN crystals tend to macro-fracture under high grinding forces. This exposes fresh, sharp cutting edges. (*Courtesy of GE Superabrasives*)

The Borazon CBN Types I, II, 500, and 510 consist of single CBN crystals (monocrystals) which have a large number of cleavage planes along which fracture occurs. This macrofracture (large break) is essential if the grains are to resharpen themselves when they become dull. This macrofracture, which may occur a number of times during the life of a crystal, keeps the crystals sharp and effective in removing steel (Fig. 2-16).

Type 550. This is an extremely tough microcrystalline abrasive designed for use in the most demanding applications that require either a vitrified- or metal-bond system. As the thermal stability of Type 550 is higher than the other types of CBN abrasives, it offers some unique advantages in the manufacture and use of metal and vitrified grinding wheels (Fig. 2-17A).

Type 560. This CBN abrasive is manufactured by applying a 60 wt % nickel coating to Type 550 crystal, making the Type 560 suited solely for use in resin-bond grinding wheels (Fig. 2-17B).

Type 570. Type 570 is manufactured by applying a special surface treatment to the Type 550 crystal which tailors it specifically for use in electroplated wheels (Fig. 2-17C).

The Borazon Type 550, 560, and 570 are microcrystalline CBN superabrasives which resharpen themselves by microfracturing (very small breaks) rather

Fig. 2-15 | Cubic boron nitride 80/100 abrasive crystals. (A) CBN 500. (B) CBN 510. (*Courtesy of GE Superabrasives*)

(A)   (B)   (C)

Fig. 2-17 | Cubic boron nitride 80/100 abrasive crystals. (A) CBN 550. (B) CBN 560. (C) CBN 570. (*Courtesy of GE Superabrasives*)

Microcrystalline CBN Particle Wear Pattern

Fig. 2-18 Microcrystalline CBN particles microfracture during grinding. Since very little of the superabrasive particles is lost in the resharpening process, it promotes long wheel life. (*Courtesy of GE Superabrasives*)

than by macrofracturing. Each particle consists of thousands of micron-size CBN crystalline regions tightly bonded to each other to form a 100 percent dense particle. When the sharp edges of the individual abrasive particles start to dull, grinding forces result in cleavage at a submicron-size CBN crystalline region. This microfracturing (Fig. 2-18) continuously resharpens each superabrasive particle as it becomes dull, resulting in consistent, efficient cutting action with a minimum of frictional heating.

The various types of CBN superabrasives and their description and the type of wheel bond system each is suited for are outlined in Table 2-2.

CUBIC BORON NITRIDE SIZING

Conventional grinding wheels are available in grit size; for example, a wheel may be called a "120-grit wheel," meaning that the average size of the abrasive grain is about 120 particles per inch. This is a very loose specification. Particles larger than 120 and

much smaller than 120 make up a rather broad distribution of grain sizes. Diamond and CBN require a much tighter size standard for optimum productivity in a given application.

Diamond and CBN are available by *mesh size;* each size is designated by an upper and lower limit. A 60/80-mesh RVG-W wheel contains abrasive which passes through a 60-mesh sieve and is retained on an 80-mesh sieve. There are very specific U.S. and International specifications regarding how much can be retained by the upper sieve and can pass through the lower sieve. In the United States, these specifications are presented in ANSI B74.16 and ASTM E-11. The mesh sizes of RVG and CBN products are presented in Table 2-3. The size refers to the abrasive grain itself and does not include the thickness of the coating.

CBN Wheel Grit Sizes

Grinding wheel manufacturers generally prefer not to identify specific mesh sizes which are used in the manufacture of their tools (Table 2-3). Very few people in the user industries are familiar with "mesh" sizes, therefore, it is only practical to continue to identify the size range of superabrasives contained in a given wheel by a "grit" size.

There is no abrasive industry standard by which mesh sizes are converted to grit sizes. The wheel manufacturers can set up their own specific designations of mesh size to grit size, but the user can rely on the fact that the relationship between mesh and grit is always very close for all manufacturers. Therefore, an 80-grit CBN wheel from any manufacturer is *coarse* and probably contains mostly 60/80 and/or 80/100

Table 2-2   BORAZON CBN TYPES

Abrasive Type	Description	Recommended Wheel Type
Type I	Basic CBN crystal, black color, Knoop hardness 4700, medium friability	Vitreous-bond and electroplated wheels; sometimes used in metal-bond wheels
Type II	Same as Type I except that crystal is coated with nickel (60% by weight); most widely used	Resin-bond wheels
Type 500	A tougher CBN crystal than Type I; golden color, blocky shape	Production-type electroplated wheels
Type 510	A CBN crystal that is coated to hold better in metal and vitreous-bond systems	Metal-bond and vitreous-bond wheels
Type 550	A microcrystalline CBN particle in which cleavage planes are essentially absent; extremely tough, with a high (1200°C) resistance to thermally induced cracking	Metal-bond and vitreous-bond wheels
Type 560	Same as CBN 550 except with 60% nickel coating; used in resin-bond systems that require high stock removal and good finishes	Resin-bond wheels
Type 570	Same as CBN 550 except treated for electroplated wheel applications	Electroplated wheels

Table 2-3 **MESH SIZES OF SUPERABRASIVE PRODUCTS**

| Mesh Size | Median Particle Diameter | | RVG Products Borazon Types I and II | Borazon 550, 560, 570 |
	in.	mm		
20/30	0.028	0.711		X
30/40	0.019	0.482		X
40/50	0.014	0.355		X
50/60	0.011	0.279		X
60/80	0.008	0.203	X	X
80/100	0.0065	0.165	X	X
100/120	0.0055	0.139	X	X
120/140	0.0045	0.114	X	X
140/170	0.0037	0.094	X	X
170/200	0.0032	0.081	X	X
200/230	0.0027	0.068	X	X
230/270	0.0023	0.058	X	
270/325	0.0019	0.048	X	
325/400	0.0016	0.040	X	

mesh CBN abrasive. A 150-grit wheel would likely contain mostly 140/170 mesh CBN abrasive, etc.

This text will always refer to the appropriate grit size of CBN wheel to be used in any application. The only exceptions in this text are in the detailed descriptions of certain case histories where a specific CBN mesh size was used and is so designated. CBN wheels are only sold according to their grit size, and grit size is the only terminology the student need use.

Micron Powders

When fine particles (50 microns or less) are used for polishing applications, mesh sizing is not sufficient to ensure that no oversize particles ruin the surface finish of a workpiece. A closer specification for sizing is *grade* used for micron powders. The grade designation specifies the mean (average) particle size. The Department of Commerce specifications, CS 261-63, defines a standard deviation on the particle size distribution and an absolute maximum permissible size.

Both diamond and CBN are available as micron powders in a variety of grades. Typical uses are as loose abrasives for metal polishing. Resin-bond wheels for polishing use the coated diamond or CBN micron powders.

Modern high-strength and hard alloys used in today's industrial applications require improved abrasives for grinding. RVG diamond and Borazon CBN fulfill this need. These products are manufactured using unique high-pressure, high-temperature processes. By proper selection of the ingredients and processing conditions, superabrasives can be made with properties specific to an application. The usefulness of the abrasive grains in bond systems is enhanced by surface treatments and coatings for better retention. A complete selection of superabrasive products is available for nearly any grinding or polishing requirement.

REVIEW QUESTIONS

1. List five advantages of high-efficiency superabrasive cutting tools.

Properties of Superabrasives

2. Name and briefly describe three important properties of superabrasives.
3. For what purpose are manufactured diamond and CBN used?

Carbon, Graphite, and Diamond

4. Name the three common forms of carbon.
5. Describe the atom arrangement in graphite and diamond.

Development of the Diamond Process

6. Why was it important that the United States develop a process for manufacturing diamond?
7. Name and describe the apparatus which was designed to manufacture diamond.
8. What was the purpose of a solvent-catalyst? Name three common metals used for this purpose.
9. Describe the RVG diamond crystal.
10. What is polycrystallinity, and why is it important?

Types of RVG Abrasive

11. Name three purposes of coating RVG crystals.
12. What coating is most commonly used and explain 56 wt % coating.
13. Define RVG-W and RVG-D.

Diamond and Steel Grinding

14. On what type of workpiece materials can diamond be used effectively?

15. Why is diamond not recommended for grinding ferrous metals?

Types of Cubic Boron Nitride

16. Briefly describe and state the purpose of the following abrasive types:
    a. CBN Type I
    b. CBN Type II
    c. CBN 500
    d. CBN 550

CHAPTER THREE

Grindability and Machinability of Metals

The development of new alloys and superalloys has made it more important than ever for the machine-tool operator and the tool and manufacturing engineer to understand the physical metallurgy of the work material. Chemical composition, tensile properties, thermal properties, microstructure, abrasiveness, and shear properties all affect how easy or difficult it will be to machine the work material. Since these factors have an effect on cutting tool selection, speeds, feeds, depth of cut, cutting-tool life, coolant, and other variables, a thorough understanding can affect the efficiency of a machining operation and the quality of product.

OBJECTIVES

After completing this chapter you should be familiar with:

1. The definition of machinability and grindability
2. The mechanical properties of metals
3. The alloying elements which influence machinability and grindability
4. The specific types of workpieces most effectively machined and/or ground with super-abrasives

MACHINABILITY AND GRINDABILITY

Machinability and *grindability* are very similar terms which define the relative ease with which a metal can be turned, milled, drilled, reamed, slotted, or ground. In a practical sense the term *relative ease* refers to the amount of tool wear and the amount of power required to machine or grind a metal. In a metal with low machinability or grindability, tool wear is generally high, tool changes are frequent, and tool costs are high. Examples are almost any of the tool steels, hardened alloy steels, and the high-temperature alloys used in aircraft engines. By contrast, low-carbon steels, soft cast irons, bronzes, and aluminum alloys are very easy to machine, and in these materials tool life is long, tool changes are infrequent, and tool costs are low.

Table 3-1 illustrates the results of a comprehensive study made to determine the machinability and con-sequently the costs of machining a broad range of workpiece materials. It should be noted that an extremely easy to machine aluminum alloy was used as a basis of comparison. Plain carbon steels and certain aluminum alloys are the easiest and cheapest to machine. Hardened alloy steels, high-strength steels, and nickel-based alloys are the most difficult and expensive to machine.

Table 3-2 shows the power required to machine selected materials at the same speed and feed conditions. Note the increase in power as the machinability of the metal decreases.

Very powerful, rigid machines are generally required to work with metals of low machinability, while very high productivity can be achieved with moderately powered equipment when machining metals with high machinability characteristics. It is

Table 3-1 **RELATIVE MACHINING COSTS FOR HARD AND SOFT MATERIALS**

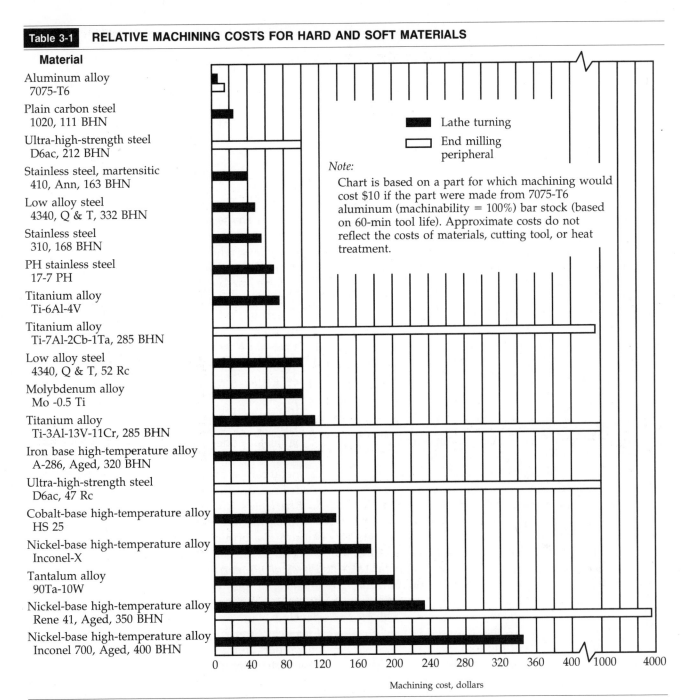

Material

Aluminum alloy
7075-T6

Plain carbon steel
1020, 111 BHN

Ultra-high-strength steel
D6ac, 212 BHN

Stainless steel, martensitic
410, Ann, 163 BHN

Low alloy steel
4340, Q & T, 332 BHN

Stainless steel
310, 168 BHN

PH stainless steel
17-7 PH

Titanium alloy
Ti-6Al-4V

Titanium alloy
Ti-7Al-2Cb-1Ta, 285 BHN

Low alloy steel
4340, Q & T, 52 Rc

Molybdenum alloy
Mo -0.5 Ti

Titanium alloy
Ti-3Al-13V-11Cr, 285 BHN

Iron base high-temperature alloy
A-286, Aged, 320 BHN

Ultra-high-strength steel
D6ac, 47 Rc

Cobalt-base high-temperature alloy
HS 25

Nickel-base high-temperature alloy
Inconel-X

Tantalum alloy
90Ta-10W

Nickel-base high-temperature alloy
Rene 41, Aged, 350 BHN

Nickel-base high-temperature alloy
Inconel 700, Aged, 400 BHN

■ Lathe turning
□ End milling peripheral

Note:
Chart is based on a part for which machining would cost $10 if the part were made from 7075-T6 aluminum (machinability = 100%) bar stock (based on 60-min tool life). Approximate costs do not reflect the costs of materials, cutting tool, or heat treatment.

0 40 80 120 160 200 240 280 320 360 400 1000 4000

Machining cost, dollars

(Courtesy of Metcut Research Associates, Inc.)

important to note that ratings or grading of machinability characteristics are based on performance of conventional tool steels, carbides, or aluminum oxide grinding wheels. The productivity potential of superabrasives may radically alter this machinability ranking, which is based on the wear characteristics of conventional tool materials.

Understanding the relative machinability and grindability of metals as well as the factors which influence these characteristics is an important aspect of understanding the total superabrasive machining and grinding processes.

Differences between Machinability and Grindability

Machinability and grindability are dependent on the same physical, mechanical, and chemical properties of a given workpiece. Generally a metal of low machinability will also be difficult to grind (DTG), and one of high machinability will also be easy to grind (ETG). But this does not mean that the difference between machinability of two workpieces and their grindability will be exactly the same. There may be slight differences in the ease or difficulty of machining or grinding which are attributable to the nature of

Table 3-2   POWER REQUIRED TO MACHINE

Material	Power required, hp*/(in.³ · min)
Aluminum alloy	1.0
Gray cast iron	1.5
Low-carbon steel	3.0
Titanium alloy	5.0
Stainless steel	7.0
High temperature alloy	7.0

*Horsepower.

the chip forming process in these two methods of metal removal.

Machining is a process of forming chips of predictable, uniform thickness and width. The thickness of the chips may range from less than 0.005 to over 0.060 in. [0.12 to over 1.5 millimeters (mm)] but are typically in the range of 0.060 in. (1.5 mm) for most machining operations. By contrast, the abrasive grains on the periphery of a grinding wheel produce extremely small chips of varying thickness, width, and shape. The typical range of thickness of chips will vary from as little as a few ten thousandths of an inch up to as much as 0.002 in. (0.002 to 0.05 mm) in coarse grinding conditions.

The chip characteristics in grinding are due to the random arrangement of abrasive grains on the surface of the grinding wheel and how far they protrude above the wheel surface along with their high negative rake characteristics. Because machining takes place by efficiently producing large uniform chips, far less power is required to remove a unit volume of workpiece than is required in grinding. Making tens of thousands of very small ships to remove a cubic inch of steel requires more energy in comparison to the energy required to produce a few dozen machining chips. In general, the specific energy required to grind away a cubic inch or cubic centimeter of a given metal will be 10 to 30 times that required to machine the same metal with a cutting tool.

The relative machinability and grindability of a given metal are determined by its properties. A discussion of which properties and how they are controlled is important.

THE PROPERTIES OF METALS

The major properties of metals can be described as follows:

1. *Mechanical Properties.* The properties of a metal that determine its behavior when external forces are applied. These are:
   a. *Tensile Strength.* The most commonly used description of strength refers to tensile strength (Fig. 3-1A). This is the capacity of a metal to withstand a "pull" or tension applied to it. However, both compression and shear strengths are closely related and of equal importance.
   b. *Fatigue Strength.* The ability of a metal to withstand millions of cycles of repeated loading without failure. Springs are an important example.
   c. *Hardness.* The ability of a metal to withstand penetration or indentation of its surface (Fig. 3-1B). This is the single most important property for determining abrasion resistance and strength.
   d. *Elasticity.* The ability of a metal to return to its original shape after being deformed by a load (Fig. 3-1C).
   e. *Toughness.* The degree to which a metal can absorb energy by plastic deformation before it fractures or breaks completely. This is a very important property for steel subjected to repeated impact shock loading such as a stamping die.
   f. *Abrasion Resistance.* The ability to resist wearing away by friction, rubbing, or grinding.
   g. *Ductility.* That property which describes the ease with which a metal or other material can be drawn or stretched into permanent deformation without rupture (Fig. 3-1D). Automobile body panels are one example.
   h. *Malleability.* The relative ease with which a metal can be hammered, forged, or rolled into a desired product shape (Fig. 3-1E).
   i. *Brittleness.* The property of a metal that permits no permanent distortion before breaking. Cast iron is a brittle metal; it will break rather than bend under shock or impact (Fig. 3-1F).
2. *Physical Properties.* Properties other than mechanical properties such as density, electrical conductivity, thermal conductivity, and thermal expansion.
3. *Chemical Properties.* The properties of metals and alloys pertaining to their reaction with their environments. Examples of this are corrosion and oxidation resistance properties.

The mechanical properties of hardness, strength, toughness, and abrasion resistance will be the main properties which determine machinability and grindability of a given metal. Physical properties will have no real effect on the relative machinability of a metal. Chemical properties can have an effect on machinability. There can be a potential chemical reaction between abrasives and cutting tools with workpieces, especially at the temperatures at which cutting and grinding take place. Therefore, the chemical composition of a workpiece can be an important element in determining its relative grindability and machinability. An outstanding example of a metal whose machinability is significantly determined by its chemical

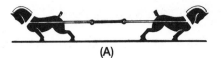

Fig. 3-1A Tensile strength is the amount that a metal will resist a direct pull. (*Courtesy of Linde Division, Union Carbide Corp.*)

Fig. 3-1B Hard metals resist penetration. (*Courtesy of Linde Division, Union Carbide Corp.*)

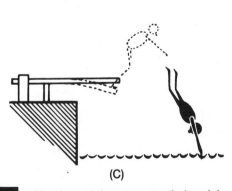

Fig. 3-1C Elastic metals return to their original shape after the load is removed. (*Courtesy of Linde Division, Union Carbide Corp.*)

Fig. 3-1D Ductile metals are easily deformed. (*Courtesy of Linde Division, Union Carbide Corp.*)

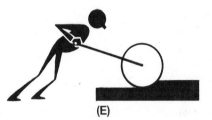

Fig. 3-1E Malleable metals are easily formed or shaped. (*Courtesy of Linde Division, Union Carbide Corp.*)

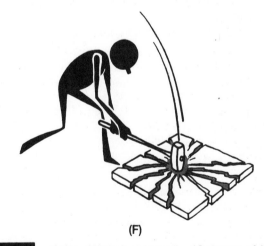

Fig. 3-1F Brittle metals break easily. (*Courtesy of Linde Division, Union Carbide Corp.*)

composition is titanium. Titanium is a highly active metal which reacts to one degree or another with virtually all types of abrasives and cutting-tool materials. This reaction typically accelerates tool wear and thus decreases machinability.

MANUFACTURE OF METALS

Iron, aluminum, and copper are among the most common metal elements found in nature. Iron, which is found in many parts of the world, was considered a rare and precious metal in ancient times. Since these ancient times, the process of alloying metals has become a tool to improve base metal properties. Alloying is the process of adding one or more other metal elements to a base metal to obtain specific properties.

Today, iron and its alloys have become the most important class of materials known to people worldwide. Steel, an alloy of iron and carbon, is of critical technical, commercial, and economic importance in everyday life.

Almost every product made today contains some steel or was manufactured by tools made of steel. Steel can be hard enough to cut other metals and glass, pliable as the steel in a paper clip, flexible as steel in a spring, or strong enough to withstand the stress in buildings and bridges. An improved understanding of the properties of this metal is of importance. A classification of steel is illustrated in Table 3-3.

The following discussion of the properties of metals pertains only to steel. However, both the discussion of properties and the influence of alloying elements on those properties will show the general metallurgical principles of alloying, especially on mechanical properties.

Physical Metallurgy

Physical metallurgy of steel is the science concerned with the production of the required physical, mechanical, and chemical properties in steel. The three major elements of physical metallurgy are:

1. *Composition.* The alloying elements will influence the mechanical properties and the response to subsequent heat treatment.
2. *Heat Treatment.* Controlled heating and cooling of steel is done to further obtain specific mechanical properties needed for the performance of the product.
3. *Mechanical Work.* Hot or cold working of steel may additionally be used to control uniformity of mechanical properties, absence of voids, amount of ductility, toughness, and grain structure.

Development of the mechanical properties of steel begins with the chemical composition used to manufacture the steel. *Chemical composition* in the case of steel refers to the various alloying elements which will be used in its manufacture. The properties of the product to be manufactured from steel, in turn, determine the alloying elements required to provide the appropriate properties. Figure 3-2 shows the steps that lead from steel product requirements to the resulting machinability and grindability characteristics.

Influence of Composition on Hardenability

Hardenability is a measure of the depth and distribution of hardness which can be obtained in a steel by heating followed by rapid cooling (quenching). The metallurgical phase or structure within the steel which is hard is called *martensite*. The hardness of martensite which can be formed is directly related to the carbon content of the steel. The hardenability of a steel is dependent on its alloying elements.

In the case of plain carbon steels, increasing carbon content, in combination with heat treatment, can

Table 3-3 CLASSES OF STEEL

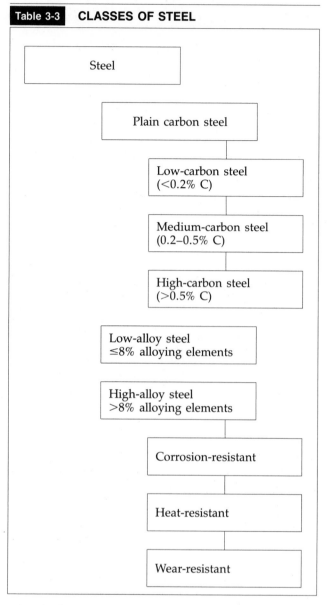

raise the base strength and hardness characteristics to higher levels. In the case of alloyed steels, a wide range of metal elements are used to create steel of almost unlimited combinations of mechanical properties. The most important elements which will relate to the hardenability, strength, and toughness are listed as follows:

Manganese	Molybdenum
Silicon	Vanadium
Nickel	Tungsten
Chromium	Cobalt

Some examples of the great variety of steels produced from these alloying elements are listed as follows:

Nickel steels	Nickel-chromium-molybdenum steels
Nickel-chromium steels	
Molybdenum steels	Nickel-molybdenum steels
Chromium-molybdenum steels	Chromium steels

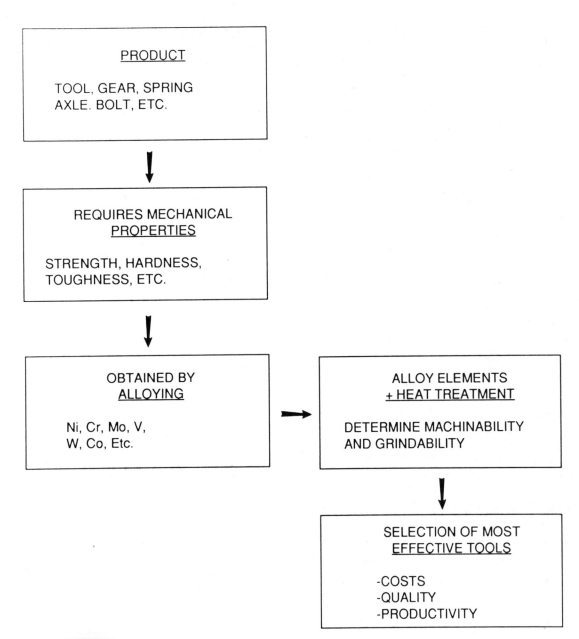

PRODUCT

TOOL, GEAR, SPRING
AXLE. BOLT, ETC.

REQUIRES MECHANICAL
PROPERTIES

STRENGTH, HARDNESS,
TOUGHNESS, ETC.

OBTAINED BY
ALLOYING

Ni, Cr, Mo, V,
W, Co, Etc.

ALLOY ELEMENTS
+ HEAT TREATMENT

DETERMINE MACHINABILITY
AND GRINDABILITY

SELECTION OF MOST
EFFECTIVE TOOLS

-COSTS
-QUALITY
-PRODUCTIVITY

Fig. 3-2 Steps from product requirements to machinability-grindability characteristics.

Figure 3-3 illustrates the difference in hardenability of a plain carbon steel bar and an alloy steel bar. Both bars have been heated to the proper temperature and then quickly cooled (quenched) in water. The maximum hardness attained in the plain low-carbon 1045 steel is only Rc 43. The chromium-molybdenum alloy 4140 bar is at Rc 60 at its surface and drops to only Rc 57 at its center. The 1045 steel has relatively low hardenability, while the 4140 has high hardenability.

In addition, some of these alloying elements have other effects in the alloying process. Chromium, molybdenum, vanadium, and tungsten have the potential for forming very hard carbides with the carbon present in the steel. Tungsten carbide, molybdenum carbide, vanadium carbide, and chromium carbide are all much harder than martensite. In high-speed

steels, tool steels, and die steels, it is desirable to use significant amounts (greater than 8 percent total) of these elements to deliberately create these extremely hard carbide inclusions within the martensite matrix. The addition of these carbides would, of course, have a considerable effect on reducing the machinability and grindability of steels.

THE MACHINABILITY AND GRINDABILITY OF OTHER IMPORTANT METALS

Cast Iron

Cast iron is another ferrous metal of great economic importance. Probably the simplest definition of cast iron is "an iron containing so much carbon that it is not malleable at any temperature." In general, cast

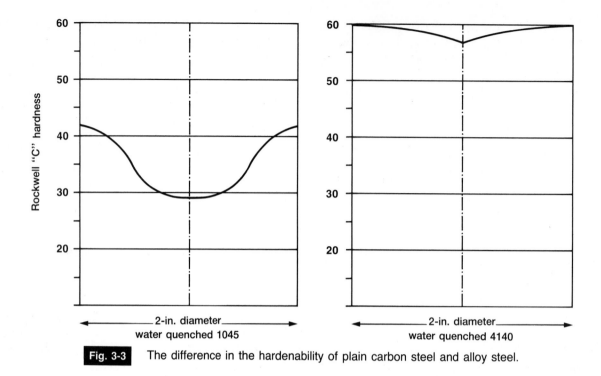

Fig. 3-3 The difference in the hardenability of plain carbon steel and alloy steel.

iron is very brittle in comparison to steel. Once cast, it can only be ground or machined to some subsequent shape. It cannot be rolled, drawn, forged, or subjected to other types of forming as steel can.

Carbon is the single most important component in the manufacture of cast iron. The distribution of carbon in the form of graphite and the various ways in which carbon can alloy with iron and other elements determine the mechanical properties of the iron and consequently its relative ease or difficulty to machine.

Three types of cast iron are of major commercial interest.

GRAY CAST IRON. Gray cast iron is very widely used in the manufacture of automobile engines, gears, and other products as well as in the machine-tool industry. Silicon is a very important component used in the manufacture of gray iron. Gray iron, like other cast irons, is a complex metallurgical structure. Two important phases which develop in gray cast iron are important in determining the machinability of a gray cast iron. While ferritic cast iron is very soft and thus easy to machine, pearlitic cast iron is much harder and is more difficult to machine.

WHITE CAST IRON. White cast irons may contain larger quantities of chromium, nickel, and vanadium as alloying elements. This produces an iron that may be

extremely hard at its surface (up to Rc 60). Steel mill rolls and other applications requiring a high degree of hardness and abrasion resistance are typical examples of the application of white cast iron. In general, white cast iron has low machinability.

DUCTILE CAST IRON. Ductile cast irons have a much higher degree of ductility than do other forms of cast iron. This, again, is achieved through the appropriate use of carbon in the form of graphite and other alloying elements such as silicon to achieve a material with a degree of toughness and ductility higher than other cast irons. Applications for ductile cast iron are automotive crankshafts, camshafts, and exhaust manifolds.

Superalloys

Superalloy (high-temperature alloy) is a word which has been coined to describe alloys designed specifically for high-temperature, highly stressed applications. The major reason for the development of such alloys has been for applications in aircraft engines. Because these alloys are designed to retain a high level of strength even at very high temperatures, they are difficult to machine. They retain their strength at the high temperatures encountered in grinding and machining, so, unlike other metals, they do not become significantly easier to machine. In addition, most of these metals have a tendency to "work-harden." Work-hardening is a characteristic which allows the strength and hardness of a metal to actually increase while it is in the process of being machined or formed. This further decreases the machinability of such metals.

There are three major families of superalloys. These are as follows:

Nickel-Based Alloys. Examples are Inconel, Rene, and Waspalloy. The important alloying elements for these materials are cobalt, chromium, molybdenum, and titanium.

Cobalt-Based Alloys. Examples are Stellite, Colmonoy, and Hastalloy. The important alloying elements here are chromium, molybdenum, and nickel.

Iron-Based Alloys. Examples are A286 and Incoloy. The primary alloying elements here are chromium, molybdenum, and nickel.

Hard Facing Materials

The term *hard facing materials* refers to a very broad range of hard, abrasion-resistant, temperature-resistant materials which can be deposited as a coating on the exterior of many products. In general, a high-temperature flame spray or a plasma spray is required, but there are also cold spraying processes which can deposit these hard facing materials. Because of the basic mechanical properties of these materials after they are deposited on a surface, they are generally very difficult to machine or grind. There are two major categories of hard facing materials:

Carbide-oxide-base materials may consist of tungsten carbides, vanadium carbides, and chromium carbides, as well as aluminum oxide, chromium oxide, zirconium oxide, and other such hard carbides and oxides.

Metal-base materials consist of molybdenum, nickel, chromium, tungsten, and iron.

Aluminum Alloys

Aluminum is another metal of extreme economic importance. Aluminum would rank number three after steel and cast iron as the most widely used metal in industry and commerce today. Major users are the aircraft industry, the automotive industry, the construction industry, and the packaging industry. In comparison with all the metals discussed thus far, aluminum alloys are in general extremely easy to machine. Certain alloys, however, become particularly difficult to machine. These are as follows:

Sand or Permanent Mold Cast. The single most important alloying element in aluminum made for this type of casting is the presence of significant quantities of silicon. Silicon may be included in these types of alloys in the range 2 to 3 percent to over 17 percent. Because silicon is hard and abrasive, the machinability of these types of aluminum alloys is sharply decreased in direct proportion to the silicon content. The use of silicon significantly increases the strength, toughness, and abrasion-resistance of these alloys.

Die-Cast Alloys. These alloys have properties somewhat similar to those made for sand and permanent mold casting.

Wrought Aluminum Alloys. An extremely broad range of properties can be achieved in wrought aluminum alloys through the use of such elements as silicon, copper, and magnesium. These are generally extremely easy to machine. Some of them will exhibit work-hardening characteristics, and others can produce extremely difficult chip handling problems when machining at high speeds.

Cemented Tungsten Carbide

These are the hardest metals that are commercially available with current technology. In addition, these metals retain their hardness at high temperatures. Therefore, they are especially suited and most widely used as cutting-tool materials. As such, they cannot be effectively ground or machined with anything other than superabrasives. A more detailed description of cemented carbides and their machinability and grindability will be found in Chapter 8 of this text.

A summary table covering the six families of metals discussed in this section is found in Table 3-4. This table provides an overview guide to those types of metals which can be most effectively ground or machined with superabrasives. The table reflects the hardness condition for each major type of metal, provides some trade names for application examples of each, lists the major alloying elements, and indicates which type of superabrasive is appropriate for grinding and machining. Frequent reference to this table will assist the student in acquiring an understanding of the relationships between workpiece properties and their machinability and grindability.

Table 3-4 **METALS COMMONLY GROUND OR MACHINED WITH SUPERABRASIVES**

Metal Types	Hardness	Examples	Principal Alloying Elements*	Grind with		Machine with	
				CBN	Diamond	PCBN	PCD
Hardened steel							
Tool, die, high-speed steels	>50 HRc >50 HRc	A-2, D-2, M-2, M-4, O-5, T-15	Co, Cr, Mo, V, W	Yes	No	Yes	No
Alloy steels	>50 HRc	4130, 4340, 5150, 52100, 8620, 9260	Cr, Mo, Ni, V	Yes	No	Yes	No
Carbon steels	>50 HRc	1050, 1095	Mn, Si	Yes	No	Yes	No
Stainless steels							
Austenitic	—	301, 302	Cr, Ni, Mn	—†	No	Yes	No
Martensitic	>50 HRc	410, 440A	Cr	Yes	No	Yes	No
Cast iron							
Gray iron	>180 BHN	Engine blocks, flywheels, crankshafts	C, Si	Yes	No	Yes	No
White	>500 BHN	Ni-Hard (rolls)	C, Ni, Si, Cr	Yes	No	Yes	No
Ductile	>200 BHN	Crankshafts, exhaust manifolds	C, Si	—†	No	—†	No
Superalloys (Hi-temp alloys)							
Nickel base	>35 HRc	Inconel, Rene, Waspalloy	Cr, Co, Mo, W, Ti	Yes	No	Yes	No
Cobalt base	>35 HRc	Stellite, Airesist, Haynes Alloy	Cr, W	Yes	No	Yes	No
Iron base	>35 HRc	A-286, Incoloy	Cr, Ni, Mo	Yes	No	Yes	No
Hard facing materials							
Carbide-oxide base	>50 HRc	UCAR LA-2, LC-4, LW-1	Al_2O_3, Cr_2O_3, WC	No	Yes	No	Yes
Metal base	>60 HRc	Stellite, Hastalloy	Mo, Ni, Cr, Co, Fe	Yes	No	Yes	No
Aluminum alloys							
Sand or permanent mold cast		A356, A390	Si	No	No	No	Yes
Die cast		A360, 390	Si, Cu, Zn	No	No	No	Yes
Wrought		2218, 7049	Cu, Zn, Mg	No	No	No	Yes
Cemented tungsten carbide							
All tool and die grades	84–95 HRa		TaC, TiC, Co	No	Yes	No	No
Presintered—all tool and die grades	—		TaC, TiC, Co	No	Yes	No	Yes
Sintered die grades	<88 HRa only		>11% Co	No	Yes	No	Yes

Adapted from: American Society of Metals Handbook.

* *Abbreviations:* Al—aluminum, Al_2O_3—aluminum oxide, Co—cobalt, Cr—chromium, Cr_2O_3—chromium oxide, C—carbon, Mn—manganese, Mo—molybdenum, MoC—molybdenum carbide, Ni—nickel, Si—silicon, Ta—tantalum, TaC—tantalum carbide, Ti—titanium, TiC—titanium carbide, V—vanadium, VC—vanadium carbide, W—tungsten, WC—tungsten carbide, ZrO_2—zirconium oxide.

†Can be machined or ground if the equipment and operating conditions are suitable for superabrasives.

1. Name six factors which can affect the machinability of a metal.

Grindability and Machinability of Metals

2. Define the term *relative ease* as it refers to machinability and grindability.
3. How does a metal with low machinability affect the cutting tool?
4. How does the machinability of a metal affect the power required for machining?
5. What type of machines are required to work with metals of low machinability?
6. What types of chips are produced by
   a. machining?
   b. grinding?
7. Compare the power requirements for machining and grinding.

The Properties of Metals

8. List the four main mechanical properties of metals which determine machinability and grindability.

Manufacture of Metals

9. What is the purpose of adding alloys to a base metal?

Physical Metallurgy

10. Name the three major elements of physical metallurgy.
11. What determines the hardness of martensite which can be formed in a metal?
12. List the eight most important elements which relate to the hardenability, strength, and toughness of a metal.
13. What alloying elements form hard carbides in steel?

The Machinability and Grindability of Other Important Metals

14. Name and give the main use for three major types of cast iron.
15. Name the three main classes of superalloys and give the major alloying elements for each.
16. Name the two major categories of hard facing materials and state what each might consist of.
17. Name three important aluminum alloys.
18. What are the hardest metals that are commercially available with current technology?

CHAPTER FOUR

Cubic Boron Nitride Grinding Wheels

In a typical machining operation, the grinding wheel is the interface between the machine and the workpiece. For the wheel to transfer energy and efficiency from the machine into the workpiece, the cutting tool must be harder than the workpiece and must also be wear-resistant. Cubic boron nitride (CBN) crystals have met and surpassed these requirements, and this superior abrasive product has been used to manufacture grinding wheels for the high-efficiency grinding of difficult-to-grind (DTG) ferrous materials.

OBJECTIVES

After completing this chapter you should be able to:

1. Know the types of wheel bond systems which are available
2. Select the proper CBN wheel type and bond systems for each grinding application

CUBIC BORON NITRIDE WHEELS

Cubic boron nitride grinding wheels have finally been recognized as superior cutting tools for grinding difficult-to-machine metals. From their initial use in toolrooms and cutter-grinding applications, CBN wheels are really making their presence felt in production grinding operations where the alternative has been to use less costly conventional abrasives which wear out at much faster rates. The applications for CBN wheels range from the elementary regrinding of high-speed cutting tools to the ultra-high-speed grinding of hardened-steel components in the automotive industries.

Cubic boron nitride grinding wheels have more than twice the hardness of conventional abrasives for grinding DTG ferrous metals (Fig. 4-1). Hardness in an abrasive is meaningless if the abrasive is too brittle to withstand the machining pressures and the heat of production grinding. The CBN abrasive crystal has the toughness to match its hardness so that its cutting edges stay sharp longer with much slower wear rates than that of conventional abrasives.

On DTG materials, conventional grinding wheels dull quickly and as a result generate high frictional heat. As the abrasive grains dull, the material-removal rate falls and it is difficult to maintain part accuracy and geometry. The CBN wheel's prolonged cutting capacity and high thermal conductivity help prevent uncontrolled heat buildup and therefore reduce the chances of wheel glazing and workpiece metallurgical damage (Fig. 4-2). The CBN abrasive is also thermally and chemically stable at temperatures over 1832°F (1000°C)—well above the temperatures generally reached in grinding. This means reduced grinding-wheel wear, with an easier task of producing precision workpiece geometry and accuracy.

CBN WHEEL MANUFACTURE

Grinding wheels, the most important products manufactured from abrasives, are composed of abrasive grains held together by a suitable bond material. The main purpose of grinding wheels in the metalworking industry are:

1. Generation of flat, cylindrical, and formed surfaces
2. Removal of stock from generally hard, difficult-to-grind metals
3. Production of good surface finishes

For grinding wheels to perform properly, they must be hard and tough to withstand the grinding

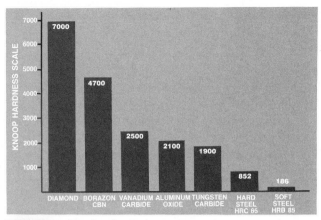

Fig. 4-1   The hardness comparison of various metals and abrasives. *(Courtesy of GE Superabrasives)*

pressures, and the abrasive grain must be capable of gradually breaking down to expose sharp new cutting edges to the material being ground. The main components of a grinding wheel are the abrasive grain and the bond material which holds them together.

Abrasive Grain

As discussed in Chapter 2, there are seven different types of General Electric (GE) CBN abrasives available, each designed to give the best performance for specific types of wheels and bond systems. These types are divided into three categories as:

1. Medium toughness, single-crystal (monocrystalline) abrasive
2. High toughness, single-crystal (monocrystalline) abrasive
3. Tough microcrystalline abrasive

The abrasive grains are also available in different mesh sizes to suit various grinding applications. They

Fig. 4-2   The part on the left, ground with an aluminum oxide wheel, shows metallurgical damage, while the part on the right, ground with a CBN wheel, shows no damage. *(Courtesy of GE Superabrasives)*

range in size from coarse grains to micron powder (20 to 800 grit), all designed for specific stock removal operations.

Cubic boron nitride wheel bonding systems fall into two categories: impregnated-bond wheels and electroplated wheels.

Impregnated-Bond Wheels (Fig. 4-3A) have CBN abrasive mixed with bond (resin, metal, or vitrified) throughout the thickness of the rim section. The bond materials must have properties to match the abrasive properties so they are designed to wear at about the same rate as the abrasive.

Electroplated CBN Wheels (Fig. 4-3B) have a single layer of abrasive which is bonded to the machined surface of a metal wheel core in a nickel electroplating bath. The electroplated nickel matrix provides outstanding retention of the abrasive.

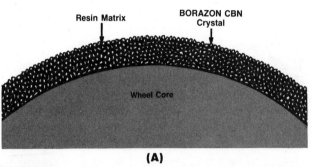

(A)

Fig. 4-3A   Impregnated-bond CBN wheels have abrasive which is mixed with bond throughout the thickness of the wheel rim. *(Courtesy of GE Superabrasives)*

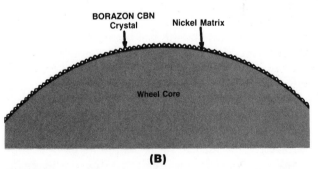

(B)

Fig. 4-3B   Electroplated CBN wheels have only one layer of abrasive bonded to the wheel core. *(Courtesy of GE Superabrasives)*

Concentration

Concentration refers to the amount of superabrasive contained in the wheel rim. A wheel with 100 concentration [72 carats per cubic inch (72 carats/in.^{3}) or 4.4 carats per cubic centimeter (4.4 carats/cm^3)] means that it contains 25 percent by volume of superabrasive. A 50-concentration wheel contains half of a 100-concentration wheel or 12.5 volume percent (vol %). *Concentration of contained abrasive* is the terminology normally used when specifying CBN wheels. See Table 4-1 for the abrasive concentrations in impregnated-bond CBN wheels. Concentration is an important factor which affects tool performance. Generally, as concentration is increased, tool life will increase. The following lists a few general guidelines regarding the effects of abrasive concentration in a superabrasive wheel.

High Concentration

- Tool life increases
- Less force per abrasive particle
- Improved surface finish

Low Concentration

- Tool life decreases
- More force per abrasive particle

Bond Types

To effectively grind the large range of materials on which CBN abrasives have proved successful, a variety of bonding systems are used to hold the abrasive grains to the surface of the wheel core. The four most common bonding systems currently used by superabrasive grinding wheel manufacturers are resinoid, vitreous, metal, and electroplated. It is important to understand how each wheel type is manufactured and the effect that the manufacturing process has on the wheel properties. These properties determine the wheel truing and dressing procedure, which is possibly the most important factor that determines whether a wheel will be effective.

RESIN-BONDED WHEELS. Resin-bonded wheels are manufactured by mixing measured amounts of phenolic or polyimid resin and filling agents with the appropriate weight and grain size of metal-coated CBN abrasive. This mixture is then used to fill the mold cavity and forms the grinding rim section around the wheel core (Fig. 4-4). After the cavity has been filled, the mixture of resin and abrasive is subjected to pressure and temperatures up to 750°F (400°C), which bonds the abrasive to the wheel core. After the wheel is removed from the mold, a further temperature curing cycle polymerizes and strengthens the resin bonding.

Depending on the grinding application, a resin-bonded wheel may contain as little as 50 concentration of CBN abrasive grains and up to 125 concentration of abrasive. Most resin-bonded CBN wheels contain from 75 to 100 concentration of abrasive. Resin-bonded wheels have generally the softest, easiest grinding action of the four types of CBN bonding

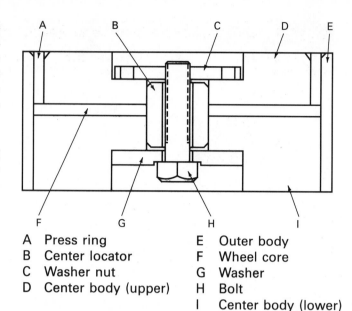

A Press ring   E Outer body
B Center locator   F Wheel core
C Washer nut   G Washer
D Center body (upper) H Bolt
                       I Center body (lower)

Fig. 4-4 The prepared wheel core in the mold ready for loading the rim mix. *(Courtesy of GE Superabrasives)*

systems. A 100 to 125 concentration of abrasive is all that the relatively weak resin bond can support.

Resin-bonded wheels have good free-cutting qualities. They can remove material fast but have limited form holding and life characteristics. CBN resin-bonded wheels are widely used in toolrooms and production applications for the grinding of tool steels, other hardened steels, and superalloys.

The key characteristics of resin-bonded CBN grinding wheels are:

- Can be effectively used in the broadest range of applications
- Available in a large range of shapes and sizes
- A fair degree of mechanical bonding
- Fairly good abrasion resistance
- Can be designed for both wet and dry grinding modes
- Good free-cutting qualities

VITREOUS-BONDED WHEELS. Vitreous-bonded wheels, also known as *ceramic-bonded wheels,* provide higher bonding strength than do resin bonds and also enable the manufacturer to vary the basic strength and chip clearance by altering the porosity and structure of the CBN wheel. Vitreous bonds provide enough bonding strength for manufacture of wheels from 50 to 200 concentration (12.5 to 50 vol % CBN abrasive). The flexibility of the vitreous-bond system makes it possible to condition the surface of this type wheel to

Table 4-1 **ABRASIVE CONCENTRATIONS IN IMPREGNATED-BOND CBN WHEELS**

Concentration Number	Percent Superabrasive (by Volume)	Superabrasive carats/in.^{3}	Superabrasive carats/cm^3	Wheel Bond
200	50	144	8.8	Vitreous
150	37.5	108	6.6	Vitreous
100	25	72	4.4	Resin, metal, vitreous
75	18.75	54	3.3	Resin, metal
50	12.5	36	2.2	Metal
25	6.25	18	1.1	Metal (rarely used)

(Courtesy of GE Superabrasives.)

achieve a broad range of metal removal rates (MRR) and surface finish characteristics. These are all reasons why vitreous bonds are becoming more important for CBN abrasives in production grinding applications.

The process for the manufacture of vitrified-bond wheels consists of the following steps:

1. The proper amount of powdered vitreous material, known as *frit*, is mixed with the desired type, weight, and grain size of CBN abrasive.
2. This mixture is carefully placed into the mold cavity which surrounds a wheel core made of a ceramic material capable of withstanding temperatures as high as 1832°F (1000°C).
3. The "green wheel" is then carefully removed and placed in a furnace with precisely time-controlled temperature and atmosphere conditions.
4. During the furnacing cycle, a chemical action takes place which causes the bonding material to form a hard, vitreous bond to hold the CBN abrasive to the wheel core.
5. Large vitrified wheels may be made of small individual radial segments which are carefully fitted and cemented to the periphery of the wheel core.

These wheels are free-cutting, produce good surface finishes, have good wear resistance, and retain straightness and form very well. The porosity (pores) or open structure of vitrified grinding wheels can be controlled to provide chip clearance, allow coolant in, and prevent wheel loading. They can be crush and rotary diamond dresser formed and are used for creep-feed grinding of hardened steel.

The key characteristics of a vitrified-bond superabrasive grinding wheel are:

- Very good abrasion resistance
- Good form-holding capabilities
- Long life
- Generally easier to true and dress than other systems
- Can generate a broader range of surface finishes
- Concentrations of 150 to 200 can produce best finishes
- Least resistance to damage due to mishandling and setup

METAL-BONDED WHEELS. Metal-bonded wheels constitute the smallest percentage of CBN grinding wheels used today. However, they have important advantages in specific areas. As metal is the toughest bonding material used in CBN wheel manufacture, these wheels are very successful in the accurate, predictable, and—of prime importance—economical stock removal of steels and superalloys.

The process for the manufacture of metal-bonded CBN grinding wheels consists of the following steps:

1. Measured amounts of powdered metals, typically a bronze or a bronze derivative, are mixed with the appropriate type, weight, and size of CBN abrasive.
2. Metal-bonded wheels generally contain from 50 to 100 concentration of CBN abrasive.
3. This mixture is carefully placed in the mold cavity which surrounds the high-temperature wheel core.
4. This entire assembly is placed in a precise temperature- and atmosphere-controlled furnace.
5. During the furnacing cycle, the powdered metal partially melts and fuses to form a very high strength bond to hold the CBN abrasive.

Cubic boron nitride metal-bonded wheels are widely used for internal, form, and creep-feed grinding applications.

The key characteristics of metal-bonded grinding wheels are:

- The bond is tough, with high abrasion and heat resistance.
- It is excellent for interrupted grinding applications.
- The wheels have very long tool life.
- Much higher power is required.
- The wheels have excellent form-holding capabilities.

ELECTROPLATED WHEELS. Electroplated wheels consist of a single layer of CBN superabrasive particles which have been bonded to the wheel surface in a nickel electroplating bath. Nickel is the most commonly used metal because it has good plating qualities and provides a bond of excellent strength. This bonding process makes it relatively easy to produce wheels of any form or contour, depending on the shape and size of the steel core. Electroplated wheels are very free-cutting, which makes high stock-removal rates possible. They are especially valuable for grinding deep forms such as gear teeth, splines, and deep grooves.

The key characteristics of electroplated superabrasive grinding wheels are:

- Lowest-cost tool form
- Complex forms easily manufactured
- Good form-holding capability

Table 4-2   GRINDING PERFORMANCE AND SPECIFICATIONS OF BOND SYSTEMS

	Wheel Bond Type			
	Resin	**Metal**	**Vitreous**	**Electroplated**
Rim depth	≥0.079 in. (2 mm)	≥0.079 in. (2 mm)	≥0.079 in. (2 mm)	One abrasive layer
Wheel life	Limited	Long	Long	Limited
G-ratio	Medium	High	Medium	N/A
Cutting action	Good to excellent	Good	Excellent	Good to excellent
Metal-removal rate	High	Limited	Medium to high	High
Form holding	Good	Excellent	Good	Good
Precautions for use	None	Rigid machines recommended	Least resistance to damage	None
Grinding	Wet or dry	Wet	Wet	Wet preferred; can be used dry

- Maximum abrasive particle exposure
- Highest stock-removal capability
- Limited finish capabilities and shorter wheel life

Grinding Characteristics and Performance

The grinding characteristics and performance of various wheel bond systems are summarized in Table 4-2.

NOTE Because there is such a variety of applications for each wheel type, these recommendations may vary somewhat for the material being ground and the grinding conditions.

WHEEL SHAPES AND IDENTIFICATION

Many shapes and sizes of superabrasive grinding wheels are produced to suit specific grinding equip-

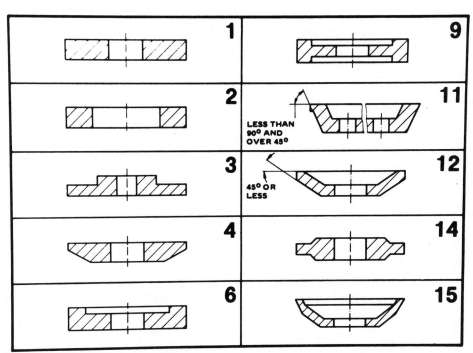

Fig. 4-5 CBN grinding wheels are available in a wide variety of shapes and sizes to suit specific grinding operations. *(Courtesy of ANSI)*

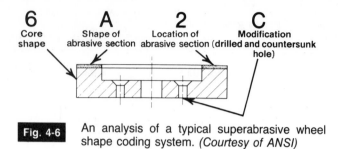

Fig. 4-6 An analysis of a typical superabrasive wheel shape coding system. *(Courtesy of ANSI)*

ment and applications. Nine basic wheel shapes are available in a range of standardized dimensional sizes (Fig. 4-5). This group of wheels is most frequently used in industrial grinding operations. The standardized wheel shapes are designated by specific type numbers which have been set by the American National Standards Institute (ANSI) and are closely followed by superabrasive wheel manufacturers. Besides the nine basic shapes, wheels are manufactured in a variation of these shapes and sizes to suit specific grinding operations and equipment.

The identification code on a superabrasive grinding wheel identifies four characteristics such as the basic core shape, abrasive section shape, abrasive section location, and wheel modifications (Fig. 4-6). This identification code is self-explanatory and is in use in the United States (ANSI) and in Europe [Fédération Européenne des Fabricants de Produits Abrasifs (FEPA)]. The location of the superabrasive

Periphery

Side

Both Sides

Inside Bevel or Arc

Outside Bevel or Arc

Part of Periphery

Part of Side

Corner

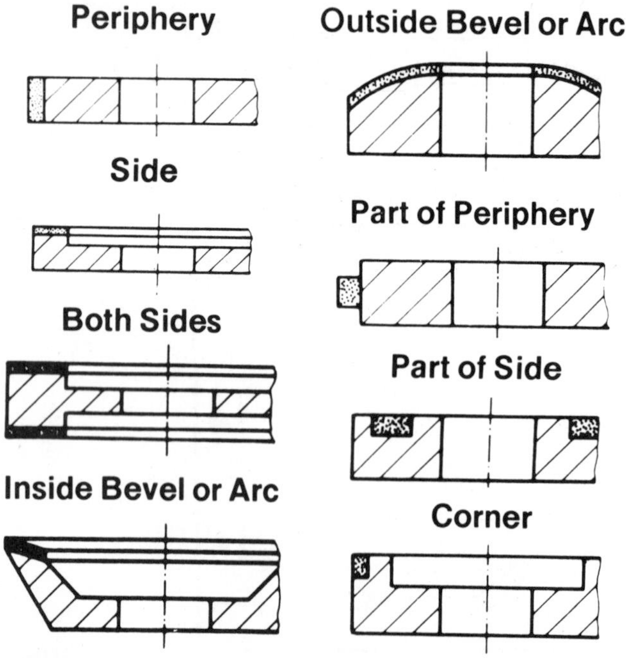

Fig. 4-7 Various CBN grinding-wheel shapes showing the abrasive section. *(Courtesy of ANSI)*

section is always on the grinding face of the wheel and a few examples of this are shown in Fig. 4-7.

Wheel Identification

United States grinding-wheel manufacturers generally use the standard ANSI marking system for superabrasive grinding wheels. This standard establishes a symbol for each of the main characteristics of a superabrasive grinding wheel and arranges these symbols in a uniform sequence. This marking system is very similar to the marking system used for conventional grinding wheels. Figure 4-8 shows and explains each part of a typical superabrasive wheel marking. This marking not only identifies a wheel in complete detail but is also useful when it is required to order a similar grinding wheel in the future.

Most European wheel manufacturers use the FEPA standard marking system to identify superabrasive grinding wheels. Figure 4-9 shows and defines a typical superabrasive grinding wheel marking using the FEPA standard.

WHEEL SELECTION

Any successful grinding operation depends to a large extent on choosing the right wheel for the job. The type of wheel selected and how it is used will affect the metal-removal rate (MRR) and the life of a grinding wheel. The selection of a CBN grinding wheel can be a complex task, and it is always wise to follow the manufacturer's suggestions for each type of wheel. They have had successful experience in designing and applying wheels for specific jobs and therefore their suggestions usually result in selecting the best wheel for each job.

Cubic boron nitride wheel selection is affected by the following factors:

- Type of grinding operation
- Grinding conditions
- Surface finish requirements
- Shape and size of the workpiece
- Type of workpiece material

All four types of CBN wheels (resin, vitrified, metal, and electroplated) are highly effective; however, they are designed for specific applications and must be selected accordingly. There is no one type of CBN wheel that is suitable for all grinding operations; therefore, for the best grinding performance, the characteristics of the abrasive must be matched to the requirements of the specific grinding job. While abrasive characteristics such as concentration, size, and

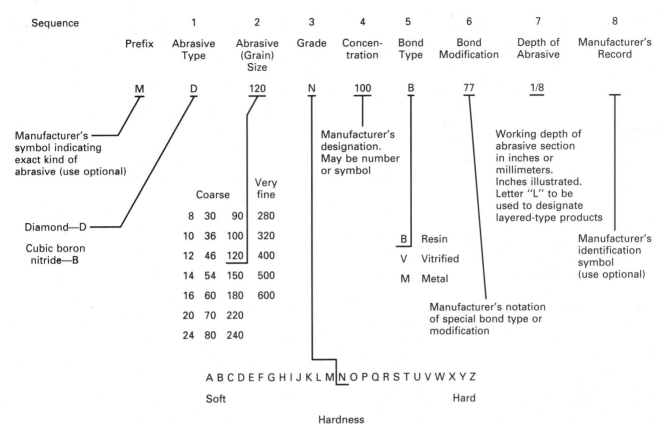

The ANSI marking system elements:

Sequence		1	2	3	4	5	6	7	8
	Prefix	Abrasive Type	Abrasive (Grain) Size	Grade	Concentration	Bond Type	Bond Modification	Depth of Abrasive	Manufacturer's Record
	M	D	120	N	100	B	77	1/8	

Manufacturer's symbol indicating exact kind of abrasive (use optional)

Diamond—D

Cubic boron nitride—B

			Very fine
Coarse			
8	30	90	280
10	36	100	320
12	46	120	400
14	54	150	500
16	60	180	600
20	70	220	
24	80	240	

Manufacturer's designation. May be number or symbol

B Resin
V Vitrified
M Metal

Manufacturer's notation of special bond type or modification

Working depth of abrasive section in inches or millimeters. Inches illustrated. Letter "L" to be used to designate layered-type products

Manufacturer's identification symbol (use optional)

A B C D E F G H I J K L M N O P Q R S T U V W X Y Z

Soft   Hard

Hardness

Fig 4-8 The ANSI marking system for diamond and CBN grinding wheels.

toughness must be considered when selecting a wheel because they affect metal removal rates, wheel life, and surface finish of any grinding operation, the wheel manufacturer understands the type of CBN abrasive available and will provide the most suitable type.

ABRASIVE CONCENTRATION. The performance of impregnated-bond wheels is strongly influenced by the concentration of CBN abrasive in the bond-matrix material.

1. High abrasive concentrations often result in higher material removal rates, longer wheel life, and improved surface finishes (Fig. 4-10).

2. Grinding power requirements tend to increase as abrasive concentration increases because there is more abrasive in contact with the work.

3. Higher abrasive concentrations can be used in vitreous-bond wheels because they maintain a stronger grip on the abrasive grains than resin bonds.

4. High abrasive concentrations for small-diameter wheels used for ID grinding, provide longer wheel life and improved surface finish.

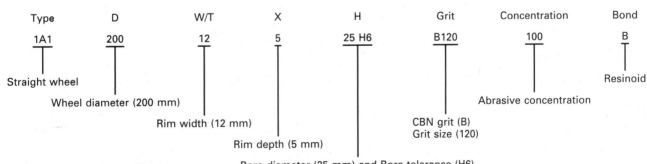

Type	D	W/T	X	H	Grit	Concentration	Bond
1A1	200	12	5	25 H6	B120	100	B

Straight wheel

Wheel diameter (200 mm)

Rim width (12 mm)

Rim depth (5 mm)

Bore diameter (25 mm) and Bore tolerance (H6)

CBN grit (B)
Grit size (120)

Abrasive concentration

Resinoid

Fig. 4-9 The FEPA marking system for diamond and CBN grinding wheels.

5. Concentration recommendations for various wheels are:
  a. 100 to 200 for vitreous-bond wheels
  b. 75 to 100 for resin-bond wheels
  c. 35 to 75 for metal-bond wheels

ABRASIVE GRIT SIZES. The size of the abrasive grain affects the metal removal rate and the type of surface finish which will be produced.

1. Coarse abrasive grain gives the wheel higher material-removal capabilities and lowers the wheel wear rate.
2. Coarse abrasive cuts a larger chip and generally produces rougher surface finishes. Microcrystalline CBN abrasive (Borazon 550, 560, and 570), which self-sharpens by microfracture, can produce smoother surface finishes than monocrystalline CBN abrasive, which self-sharpens by macrofracture.
3. When replacing an aluminum oxide grinding wheel with a CBN wheel, the abrasive grit size can be reduced by as much as three sizes without reducing the MRR, making it possible to combine the fine surface finishes with high productivity. For example, a 46-grit aluminum oxide wheel can be replaced with a 100 grit CBN wheel.

ABRASIVE TOUGHNESS. The toughness of CBN crystals or particles affects the self-sharpening characteristics of the abrasive and can have a significant effect on the performance of CBN wheels in certain grinding applications (Fig. 4-11). Although CBN crystals have a high resistance to wear, their sharp edges may eventually dull if the abrasive, bond system, and grinding conditions are mismatched. Ideally, the crystal should fracture in a controlled manner to expose fresh, sharp cutting edges.

1. If the crystal is too tough for the application, fracture may not occur. As a result, a polished wear flat is created on the crystal, which no longer cuts effectively.
2. If the crystal is not tough enough for the application, it may fracture prematurely, wasting abrasive, shortening wheel life, and generating a rougher surface finish.
3. Medium-toughness crystals should be used for most jobs.
  a. Higher-toughness crystals are recommended for operations where grinding forces are higher.
  b. Microcrystalline CBN types are recommended where grinding conditions are exceptionally severe.

Wheel Selection Guidelines

Cubic boron nitride grinding wheels are available in a complete range of shapes and sizes, such as wheels with straight or formed faces, ring wheels, disk wheels, flaring cup wheels, mounted wheels, mandrels, and hones (Fig. 4-12). Individually engineered wheels are also available to suit specific superabrasive grinding systems or specific grinding operations. As with all wheels which use high-cost abrasives, the wheel is constructed with a precision preformed core with the abrasive portion on the grinding face of the wheel (Fig. 4-7). The abrasive portion is usually between 1/16 and 1/4 in. (1.5 and 6 mm) in depth, and this information is shown on the identification label of the wheel. There generally is a CBN wheel readily available to suit any grinding operation.

The wheel manufacturer will provide the right CBN abrasive for the bond system selected. The actual way in which the wheel is used, however, will determine whether ideal wear conditions are achieved. Always be sure that the actual operating conditions are within the range of capabilities of the wheel. To be successful in the use of CBN grinding

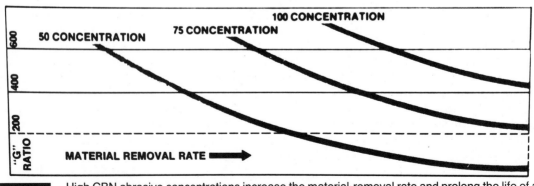

Fig. 4-10 High CBN abrasive concentrations increase the material-removal rate and prolong the life of a grinding wheel. *(Courtesy of GE Superabrasives)*

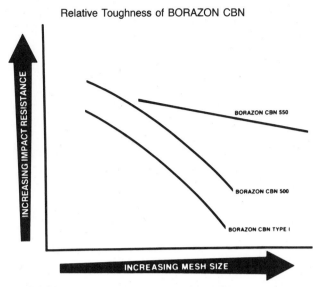

Relative Toughness of BORAZON CBN

INCREASING IMPACT RESISTANCE

BORAZON CBN 550

BORAZON CBN 500

BORAZON CBN TYPE I

INCREASING MESH SIZE

Fig. 4-11 The toughness of CBN abrasive grain can have a significant effect on its grinding performance. *(Courtesy of GE Superabrasives)*

wheels, it is important to follow these general guidelines.

1. *Select the Appropriate Bond.* Refer to bond selection in Table 4-2 as a good first choice for all major applications.
2. *Specify Normal Wheel Diameters and Widths.* When an aluminum oxide wheel is being replaced with a CBN wheel, it is recommended that the CBN wheel be of the same diameter, but the width may be reduced by 25 to 50 percent, depending on the specific nature of the application. It is not wise to use undersize wheels because they wear faster than normal-size wheels and may increase the wheel cost per ground piece. If the machine horsepower or rigidity makes it necessary to use an undersize wheel, it is better to reduce the width of a wheel rather than its diameter. If the wheel diameter is reduced, the surface speed of the wheel may be lowered to a point where the CBN wheel will not perform effectively.
3. *Choose the Largest Abrasive Mesh Size That Produces the Desired Finish.* For any given set of grinding operations, a wheel containing coarse (large mesh size) CBN abrasive will have a longer life than will a wheel which contains fine (small mesh size) abrasive. However, the wheel with fine abrasive will produce better surface finishes if the downfeed rates are lowered.
4. *Choose Wheels with the Optimum Abrasive Concentration.* While low-concentration wheels will generally do the job, they may not always be the most cost-effective. Always select the highest concentration which the grinding machine has the power to drive effectively. Concentration recom-

Fig. 4-12 CBN grinding wheels and abrasive products are available in a wide variety of shapes and sizes. *(Courtesy of Norton Co.)*

mendations for each specific grinding application will be given in the grinding section of the book.

For most surface grinding operations, resin-bond wheels with 100 concentration (containing 72 carats/in.^{3} or 4.4 carats/cm^3) of CBN abrasive on the wheel rim is recommended. On large-diameter wheels and machines with limited horsepower, it may be necessary to choose a 75-concentration CBN wheel to prevent spindle slowdown. CBN wheels with 75-concentration abrasive are very effective for sidewheel and slot-grinding operations. Table 4-1 shows the abrasive concentration in impregnated-bond CBN wheels.

Many factors can affect the efficiency of a grinding operation besides the grinding wheel. Some of the more common factors such as speed, feed, workpiece material, condition of the grinder, and cutting fluids must be considered when selecting and using a wheel so that it provides the most cost-effective operation. All these factors will be considered as they apply to specific grinding applications later in this book. Table 4-3 is a general guide which can be used for the selection of the wheel bond system.

Table 4-3   RECOMMENDED CBN BONDS FOR SPECIFIC GRINDING PROCESSES

Grinding Process	Recommended Wheel Bonds*
Surface grinding (wet)	
Horizontal spindle, reciprocating table	Resin
Horizontal spindle, rotary table	Resin
Vertical spindle, rotary table	Resin, metal
Tool and cutter grinding (dry)	Resin
Jig grinding (dry)	Electroplated (small ID's)
	Resin
	Vitreous (best finishes)
Cylindrical grinding (wet)	Resin
	Vitreous
Internal grinding (wet)	Resin
	Vitreous
Form grinding (wet)	Electroplated (deep forms)
	Resin
	Metal
	Vitreous (shallow forms)
Centerless grinding (wet)	Resin
	Vitreous
Double-disk grinding (wet)	Resin

*The wheels listed are a good first choice for the majority of grinding jobs. Other types of CBN wheels may also deliver outstanding performance in these applications, depending on the workpiece material, grinding conditions, and productivity and quality requirements.

REVIEW QUESTIONS

Cubic Boron Nitride Wheels

1. What is the range of applications of CBN wheels in the metalworking industry?
2. State four reasons why CBN wheels are superior to conventional wheels for grinding difficult-to-grind ferrous metals.

CBN Wheel Manufacture

3. List three purposes for grinding wheels in the metalworking industry.
4. Name the two categories in which CBN wheels fall.
5. Explain the difference between each wheel category in question 4.
6. What volume percent abrasive does a 100-concentration wheel contain?
7. What occurs when the abrasive concentration of a wheel increases?

8. Name four bond systems used in the manufacture of CBN grinding wheels.
9. What is the range of volume percent CBN abrasive grains in a resin-bonded wheel?
10. List three important qualities of a resin-bonded wheel.
11. What advantages do vitreous-bond wheels have over resin-bonded wheels?
12. List four key characteristics of metal-bonded CBN grinding wheels.
13. How are electroplated wheels manufactured?
14. For what purpose are electroplated wheels used?

Wheel Shapes and Identification

15. Name the standards by which grinding wheels are manufactured in the United States and Europe.

16. Explain each of the following wheel marking codes:

    B 100 J 75 M $\frac{3}{16}$

Wheel Selection

17. Name five key factors which affect CBN wheel selection.
18. List three grinding factors which are affected by abrasive concentration.
19. What grinding factors are affected by coarse abrasive grains?
20. Why is it important that the right CBN crystal toughness be selected?

Wheel Selection Guidelines

21. What type of grinding wheels are recommended for:
    a. general-purpose grinding?
    b. grinding complex forms?
22. When replacing an aluminum oxide wheel with a CBN wheel, how should the wheel sizes compare?
23. What type of wheel and wheel concentration is recommended for most surface grinding operations?
24. For what purpose are 75-concentration wheels very effective?

CHAPTER FIVE

Preparing the CBN Wheel and Grinder

Cubic boron nitride (CBN) abrasive was introduced by the General Electric Company (GE) in 1969 to solve problems related to increasing productivity and achieving higher quality in the metalworking industry. Since that time, CBN grinding wheels are continuing to replace aluminum oxide wheels for many grinding applications. Their most important uses have been on hardened alloy steels which have a low grindability factor. CBN grinding wheels have been used successfully for jig grinding, tool and cutter resharpening, surface grinding of dies, broach grinding, and the grinding of superalloys in aircraft engine manufacturing.

OBJECTIVES

After completing this chapter you should be able to understand:

1. The machine characteristics necessary to use CBN grinding wheels effectively
2. The procedure and importance of truing and dressing CBN wheels
3. The conditions which are necessary for CBN grinding wheels to reach maximum performance

GRINDING MACHINES

The performance of a CBN grinding wheel depends on the capabilities of the machine. For CBN grinding wheels to work effectively, they *must* be used on grinders which have the characteristics shown in Fig. 5-1. Trying to use superabrasive grinding wheels to make up for poor machine conditions will be doomed to failure right from the beginning.

Spindle Bearings. Loose spindle bearings will cause vibration and chatter during the machining operation shortening the grinding-wheel life, producing a poor surface finish, and inaccurate work.

Machine Slides. The slides must be close-fitting to prevent vibration and chatter which reduces the effectiveness of superabrasive grinding wheels and results in poor wheel life, poor surface finish, and inaccurate work.

Spindle Speeds. Constant spindle speeds and the ability to handle the torque required for high metal-removal rates are necessary to keep the CBN

grinding wheel operating at maximum efficiency. Loss of spindle speed reduces the efficiency of the cutting action and shortens the life of the CBN wheel.

Feeds. A reliable feed system is required to produce good surface finishes and maintain good CBN wheel life.

Coolant System. A constant supply of cutting fluid properly applied can extend the life of the superabrasive grinding wheel by as much as ten times and produce better surface finishes on the workpiece.

Before replacing an aluminum oxide grinding wheel with a CBN wheel, it is important to check the grinder to make sure that it can take advantage of the productivity potential of the superabrasive grinding wheel. Some of the factors which must be considered are horsepower, spindle speed, feed mechanism, truing and dressing system, and coolant system.

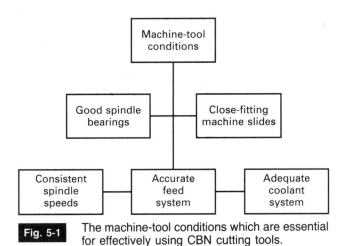

Fig. 5-1   The machine-tool conditions which are essential for effectively using CBN cutting tools.

Horsepower

A grinder must have enough power to maintain the desired wheel speed during the grinding operation. Many grinding machines in use today have enough horsepower for the conventional abrasive wheels for which they are designed. When feed rates are increased, there is enough power reserve because conventional wheels will break down as grinding pressure increases. CBN wheels, being much harder and stronger, do not break down as feed rates are increased. They will need more power to grind in order to prevent damaging slowdown or stall (Fig. 5-2).

The materials usually ground with CBN wheels are those which are hard, abrasion-resistant, and very difficult to grind. These materials will require more horsepower per unit of material removed than do more easily ground materials. For example, if it required 5 horsepower to remove ½ in.^{3}/min (8.2 cm^3) from hardened carbon steel, it might require 8 to 10

horsepower (hp) to remove the same amount from hardened M-4 or T-15 steel. The minimum and preferred spindle horsepower requirements for grinding with CBN wheels are shown in Table 5-1.

Wheel Speed

In wet grinding, higher wheel speeds can improve CBN wheel life and metal-removal rates. Excellent results are obtained with CBN wheels by grinding at wheel speeds of 5000 to 6500 surface feet per minute (sf/min) [28 to 33 meters per second (m/s)] and higher.

NOTE   Under no circumstances should a grinding wheel be operated at speeds higher than those recommended by the manufacturer.

The grinder, however, must be able to maintain *constant spindle speeds* to achieve the maximum productivity from CBN wheels. If the spindle speed slows down due to the pressure of the grinding operation, the life of the CBN will be shortened.

Figure 5-3 shows the relationship between wheel speed and its generalized effects on wheel life and stock-removal capabilities. At higher speeds, especially with large-diameter CBN wheels, it is extremely important that the wheels be dynamically balanced to produce very fine surface finishes. Table 5-2 lists the typical spindle speeds and resulting surface speeds for various sizes of CBN grinding wheels.

Feed Mechanism

An important factor about CBN wheels is that the grinder feed rate is the actual rate at which material is being removed from the workpiece. Because CBN wheels wear so slowly, the operator does not have to

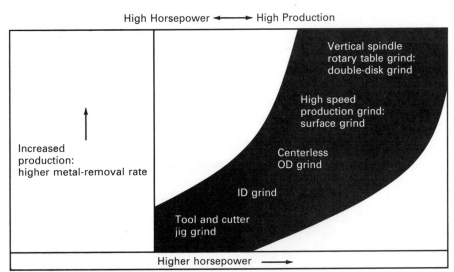

Fig. 5-2   The main advantage of using superabrasive grinding wheels is to obtain higher production rates. *(Courtesy of GE Superabrasives)*

| Table 5-1 | HORSEPOWER REQUIREMENTS FOR BORAZON CBN WHEELS |

Nominal Wheel Diameter		Horsepower per Inch (25 mm) of Wheel Width	
in.	mm	Minimum	Preferred
6–10	152–254	2	5
10–12	254–306	3	$7\frac{1}{2}$
14–16	356–406	5	15
20–24	508–610	$7\frac{1}{2}$	20
30	762	10	25

| Table 5-2 | STANDARD SURFACE GRINDER SPECIFICATIONS |

Nominal Wheel Diameter		Spindle Horsepower	Spindle Speed		
in.	mm		r/min	sf/min	m/s
7	178	1	3450	6200	31
10	254	3	2200	5800	29
12	305	5–10	1750	5500	28
14	356	5–10	1500	5500	28
14	356	5–10	1750	6400	32
20	508	$7\frac{1}{2}$–40	1200	6200	31
24	610	20–40	900	5500	28

make allowances for wheel wear when setting downfeeds. The amount that is fed is the amount that is removed from the workpiece. Grinders which use conventional abrasives are often designed to accommodate rapid wheel wear. An aluminum oxide wheel cannot be replaced with a CBN wheel without adjusting the feed rate. The low wear characteristics of CBN wheels means that they are very hard and, as a result,

produce surface finishes with characteristics somewhat different from those produced by conventional grinding wheels.

Truing System

Because CBN grinding wheels are much harder and stronger, a truing system especially designed for CBN is absolutely critical to the effective and eco-

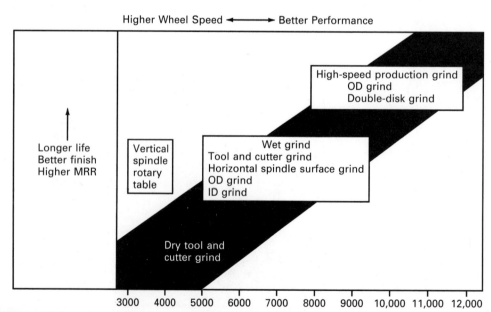

Fig. 5-3 Higher wheel speeds increase wheel life, improve material-removal rates (MRR), and produce better surface finishes. *(Courtesy of GE Superabrasives)*

nomic use of these superabrasive grinding wheels. The importance of properly truing and dressing CBN wheels cannot be overstressed; if they are not properly trued and dressed, *they will not work* effectively. The procedure for truing and dressing the various types of CBN grinding wheels will be covered in detail later in this section.

The wheel truing devices generally supplied with conventional grinding machines are not suitable for use with CBN wheels. Because of the hardness of the CBN abrasive, a flat is quickly worn on single-point or cluster-type diamonds, which are widely used with conventional grinding wheels. The most common devices used to true CBN wheels are impregnated diamond dressers and brake-controlled silicon carbide wheel dressers. For the best results on CBN wheels, it is important that the grinder be equipped with a truing device which has the necessary rigidity and is capable of feed increments of 0.0001 in. (0.002 mm).

Coolant System

The selection and application of the proper grinding fluid is a major factor in the performance of CBN wheels (Fig. 5-4). Fluids consisting of only water and a rust inhibiter have a detrimental effect on the CBN wheel performance and must not be used.

- *Light-duty soluble oils* give good results in many applications such as tool and die steel grinding operations.

- *Heavy-duty soluble oils* give even better results in these applications.

- *Sulfurized or sulfochlorinated straight oil* is gen-

erally recommended for the most difficult-to-grind materials.

- *Heavy-duty emulsifiable oil* (10 to 20 percent solutions) provide an excellent combination of lubricity and cooling capability.

When using CBN wheels on existing grinders, a higher-horsepower pump motor may be required to pump straight oil instead of a synthetic fluid. Mist collectors and fire retardation equipment may also be required for straight oils.

NOTE Specific coolant recommendations will be given for each grinding application or example as they appear later in the text.

CBN GRINDERS

The growing demand for increased productivity and higher precision in the metalworking industry has caused some machine-tool builders to design equipment specially to use CBN wheels. While CBN is effectively used on machines designed for conventional abrasives, the full potential of CBN wheels is obtained by using them on machines specifically designed to use CBN superabrasives (Fig. 5-5). These new machines are designed for better productivity and offer many advantages.

- For the equivalent production capability of conventional abrasives, the machine can usually be smaller (less floor space).

- Smaller wheels can be used (easier handling).

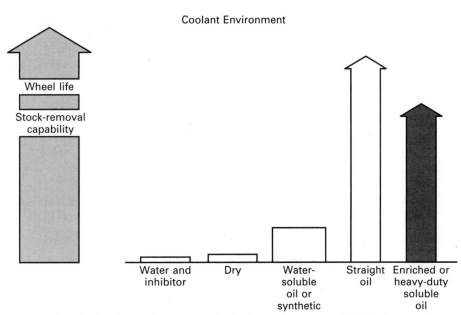

Coolant Environment

Fig. 5-4 Grinding fluids play an important role in the performance of CBN wheels. *(Courtesy of GE Superabrasives)*

The Huffman CBN grinder grinds all cutting surfaces of end mills from a solid blank with no wheel truing or dressing. *(Courtesy of GE Superabrasives)*

- Automatic operation can be improved because of extremely low wheel wear (tight tolerance capability).

- Speeds and feeds can be optimized because of constant wheel performance.

- In-process dressing is very infrequent compared to aluminum oxide wheels (more uptime).

- Wheel guarding and coolant placement can be placed in the best location and left there for the life of the wheel (with no need to design wheel size-change accommodation into the coolant system).

- The machine can be made extremely stiff as overhang or extended feed controls are kept to a minimum. A machine tool must be stiff if it is to work at high performance levels.

Production grinding machines specifically designed for CBN have five things in common:

1. They provide high productivity.
2. They use high coolant flow rates.
3. They are strong and rigid.

4. They have special automatic truing and dressing systems.
5. They have accurate programmable feed systems.

MOUNTING THE WHEEL

The importance of correctly mounting a CBN wheel cannot be overstressed. An incorrectly mounted wheel may run out of true, and as a result, too much would have to be removed from the wheel to make the grinding face true or concentric. This would be very costly since it wastes expensive abrasive and therefore reduces the wheel life.

For best results with CBN wheels, it is recommended that a high-quality standard wheel adapter (Fig. 5-6) be used when mounting a wheel and wherever possible that the wheel and adapter remain together as a unit for the life of a wheel. Whenever it is necessary, the wheel-adapter unit should be removed from the grinder. This would not only prolong the life of the wheel but also save valuable time in retruing the wheel.

The standard adapter shown in Fig. 5-6 is generally used on tool and cutter grinders and surface grinders. This adaptor has a tapered internal diameter to suit the grinder spindle taper and a parallel outside diameter to suit the hole in the grinding wheel. A solid shoulder on the adapter serves as the inner flange, while the threaded nut, which is used to tighten the grinding wheel on the adaptor, is the outer flange.

NUT   ADAPTER BODY

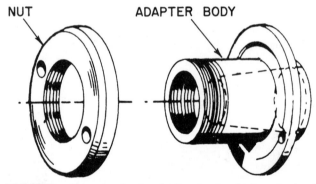

The standard spindle-wheel adapter used on tool and cutter and surface grinders. *(Courtesy of the Do All Co.)*

To Mount a CBN Wheel on an Adapter

1. Remove any burrs from the adapter, especially the solid face which serves as the inner flange.
2. Thoroughly clean the adapter and spray it with a light film of oil.
3. Mark one groove in the adaptor with a scriber or file and place a pencil mark on the wheel side (Fig. 5-7). If it should be necessary to remove the wheel from the adapter, they can be realigned

Fig. 5-7 Aligning a mark on the wheel side with a mark on the wheel adapter allows a wheel to be removed and replaced fairly close to the original position. *(Courtesy of M. Rapisarda and GE Superabrasives)*

close to the original position with a minimum of wheel runout.

4. Check the grinding-wheel sides and the internal diameter; remove any burrs and clean thoroughly.
5. Slide the wheel on the adapter; it should be a close-sliding fit and must never be forced on.
6. Set the assembled unit on a wheel assembly fixture (Fig. 5-8), and align the pencil mark on the wheel with the mark in the flange groove.
7. Hand-tighten the adapter nut to keep the wheel and adapter aligned.

Fig. 5-8 Using a wheel assembly fixture to assemble the grinding wheel and adapter. *(Courtesy of M. Rapisarda and GE Superabrasives)*

8. Hand-tighten the adapter nut securely with an adapter wrench.
9. Hold the adapter wrench in the holes of the nut with one hand and give the end of the wrench a sharp tap with a soft-faced hammer to secure the wheel adapter assembly (Fig. 5-9).

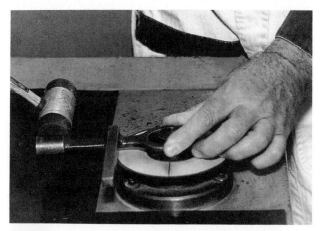

Fig. 5-9 Use a soft-faced hammer to securely tighten the wheel-adapter assembly. *(Courtesy of M. Rapisarda and GE Superabrasives)*

10. Try to turn the wheel by hand to see that the nut is tightened on the grinding wheel body and not on the screw threads of the adapter.
11. Clean the grinder spindle and flange and remove any burrs.
12. Align the mark on the wheel or the adapter with a corresponding mark on the grinder spindle (Fig. 5-10).

Fig. 5-10 Align the mark on the wheel or adapter with a corresponding mark on the grinder spindle. *(Courtesy of M. Rapisarda and GE Superabrasives)*

13. Firmly seat the wheel-adapter assembly on the grinder spindle.
14. Tighten the grinder spindle nut securely by a sharp tap on the wrench with a soft-faced hammer (Fig. 5-11).

15. Mount a dial indicator and check the wheel runout. It should not be more than 0.001 in. (0.02 mm). If the runout is more, it may be wise to recheck the assembly procedure.
16. Adjust the coolant nozzle as close as possible to the point of wheel-workpiece contact.

Mounting Large Wheels

Grinding wheels used on cylindrical and production grinders are generally larger than those used on tool, cutter, and surface grinders. They have a larger center hole which either mounts directly on a grinder spindle or on a wheel adapter. Although many mounting procedures are the same as with small wheels, there are differences in mounting large wheels.

PROCEDURE

1. Remove burrs from the diameter and shoulder of the spindle and wheel adapter (Fig. 5-12A).
2. Thoroughly clean the spindle and adapter and apply a light film of oil.
3. Remove any burrs from the faces of the wheel core and from the center hole.
4. Slide and *never force* the grinding wheel onto the grinder spindle or adapter. It should be a close-sliding fit.
5. Insert the screws in the outer flange and tighten them *lightly* in the proper sequence (Fig. 5-12B).
6. Put a dial indicator on the grinder table and set it against the circumference of the wheel (Fig. 5-13).

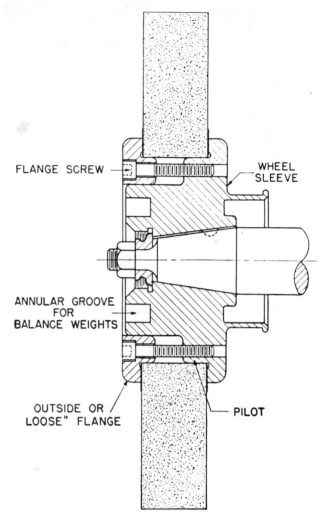

FLANGE SCREW

WHEEL SLEEVE

ANNULAR GROOVE FOR BALANCE WEIGHTS

OUTSIDE OR "LOOSE" FLANGE

PILOT

(A)

7. Have the dial indicator registering about one-quarter of a turn and then revolve the grinder spindle by hand to determine wheel runout.
8. Stop when the high spot on the grinding-wheel circumference is near the indicator.
9. Using a piece of hard rubber or soft wood, tap the high spot of the wheel away from the indicator about one-half the difference of the runout (Fig. 5-14).
10. Continue the procedure in step 9 until the runout on the wheel circumference is 0.001 in. (0.02 mm) or less.
11. Use a torque wrench of the correct size (Fig. 5-12B) and tighten all screws *in the proper sequence* to 15 foot-pounds (ft · lb). Larger wheels may require more tightening torque.
12. Recheck the trueness (concentricity) of the wheel circumference with the indicator.

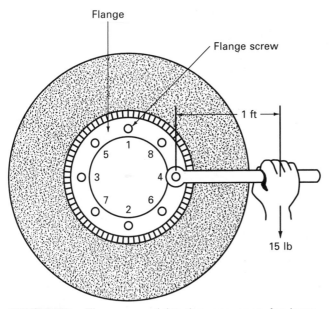

Fig. 5-12B The proper tightening sequence for large-wheel-flange screws. *(Courtesy of Grinding Wheel Institute)*

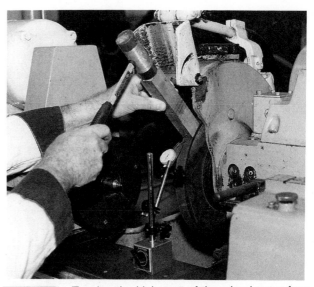

Fig. 5-14 Tapping the high spot of the wheel away from the indicator to make the wheel concentric. *(Courtesy of M. Rapisarda and GE Super-abrasives)*

PREPARING THE CBN WHEEL

Four major bonding systems are used in grinding wheels which contain CBN abrasives: resin, metal, vitreous, and electroplated. Of these types, only electroplated wheels can be used without specific wheel preparation procedures. The others, known as *impregnated bonds*, require a two- or three-step operation to true the wheel into concentricity and dress open the wheel face to expose the working crystals. Poor performance with respect to material removal rates and surface quality is the common result of an incor-

rectly prepared CBN grinding wheel. Preparing the CBN wheel may involve the operations of truing, dressing, and conditioning, depending on the wheel bond system and the grinding application.

Truing

Truing is the process of making a grinding wheel round and concentric with its spindle axis and to produce the required form or shape on the wheel (Fig. 5-15). This procedure involves the grinding or wearing away of a portion of the abrasive section of a grinding wheel to produce the desired form or shape. Successful use of the CBN wheels will require that the wheel surface has the characteristics shown in "After truing" in Fig. 5-16.

- Concentric with the spindle axis
- Free of lobes or irregularities
- Straight across for the thickness for Type 1 wheels

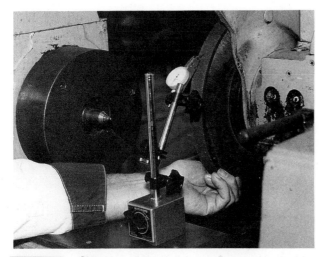

Fig. 5-13 Checking the wheel runout on a cylindrical grinder. *(Courtesy of M. Rapisarda and GE Superabrasives)*

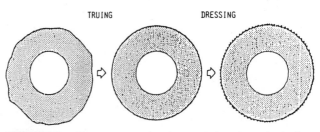

TRUING   DRESSING

Fig. 5-15 The process of truing and dressing of grinding wheels.

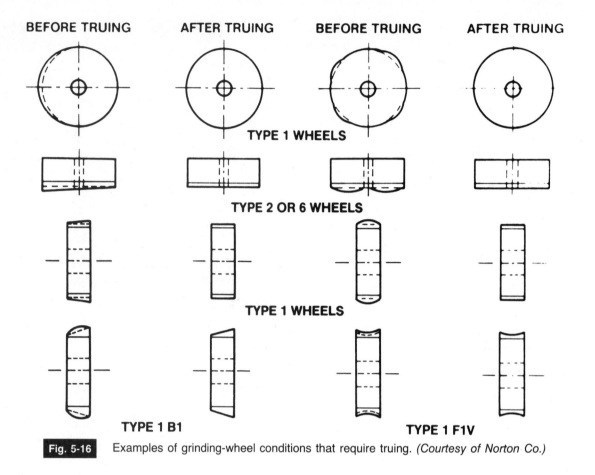

BEFORE TRUING   AFTER TRUING   BEFORE TRUING   AFTER TRUING

TYPE 1 WHEELS

TYPE 2 OR 6 WHEELS

TYPE 1 WHEELS

TYPE 1 B1   TYPE 1 F1V

Fig. 5-16 Examples of grinding-wheel conditions that require truing. *(Courtesy of Norton Co.)*

- The correct profile for form wheels
- Straight or flat on the face for Type 2 or Type 6 wheels

A properly trued wheel will grind with minimum or no chatter and produce a straight cylinder, flat surface, or accurate profile on the workpiece provided it is also dressed properly. Truing methods for CBN grinding wheels fall into three general categories: stationary tool truing, powered truing methods, and form truing (Fig. 5-17). The most common truing devices are listed in Table 5-3 along with the types of CBN grinding wheel bond system for which each is recommended.

Truing Devices and Guidelines

There is a wide variety of truing devices used for CBN grinding wheels, and generally each type works best on specific wheel bond systems. Some of the more common truing devices are described as follows:

Single-point diamond and multi-point or cluster-type diamond nibs (Fig. 5-18) are generally not recommended for resin- or metal-bond CBN wheels because a flat is quickly ground on the diamond, which reduces effective truing and produces excessive heat which could damage the wheel bond.

STATIONARY TOOL TRUING

Nib

POWERED TRUING METHODS

Rotary Cutter   Rotary Cup

FORM TRUING

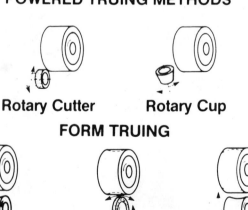

Diamond Roll   Tracing   Crush

Fig. 5-17 The types of truing methods commonly used for CBN grinding wheels. *(Courtesy of Norton Co.)*

They can be used for truing small-diameter vitreous wheels [1 in. (25 mm) or smaller] used for internal and jig grinding.

Impregnated diamond nibs (Fig. 5-19), which are generally $\frac{3}{8}$ in. (9 mm) in diameter or less to reduce frictional heat, have 100/120-mesh diamond in a metal bond matrix. They may be used to true CBN resin and vitreous bond wheels up to 8 in. (200 mm) in diameter and up to 1 in. (25 mm) wide. The diamonds in this type of tool provide a large number of cutting edges which distribute the heat over a relatively large area, and as a result, reduce excessive wear.

A brake-controlled truing device (Fig. 5-20) is the most common tool used for truing straight-faced metal and resin-bond CBN grinding wheels. This device allows effective removal of the bond matrix without creating damaging frictional heat or severe crushing of the CBN abrasive particles. Brake-controlled truing systems are designed to use an abrasive wheel such as aluminum oxide or silicon carbide, which is running at surface speeds much slower than the grinding wheel being trued. A typical abrasive wheel used in brake-controlled dressers is the vitreous bond, 60 grit, L-hardness silicon carbide wheel. This truing system is usually a gentler truing process which leaves the wheel with more texture than would a wheel trued with a diamond nib. Truing with a brake-controlled device is slower, however, and a perfectly flat wheel face is difficult to obtain because of the fast wear rate of the truing wheel.

Rotary-powered truing devices (Fig. 5-21A to 5-21C) are the most widely recommended for

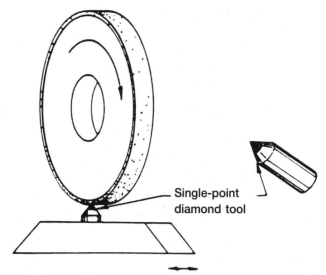

Single-point diamond tool

Fig. 5-18 Single-point diamond nibs should be used only for truing small-diameter vitreous-bond CBN wheels. *(Courtesy of GE Superabrasives)*

truing resin-, metal-, and vitreous-bond CBN wheels and are especially well suited for large-diameter wheels [8 in. (200 mm) and greater]. These devices may be pneumatic, hydraulic, or electrically powered and generally use metal-bonded or electroplated diamond truing wheels. The axis of the truing wheel may be set at an angle of 30, 45, 60, or 90°, or parallel to the axis of the grinding wheel and still provide fast and most effective truing action for both small and large wheels. Rotary-powered truing devices are most often used in production grinding operations for truing cylindrical, centerless, internal, double-disk, and large CBN grinding wheels.

Table 5-3 RECOMMENDED TRUING DEVICES FOR CBN WHEEL BOND SYSTEMS

	Wheel Bonds*			
	Resin	**Metal**	**Vitreous**	**Electroplated†**
Impregnated diamond nib	+	0	0	N/A
Single-point diamond nib	0	−	0	N/A
Rotary-powered diamond truing device	+	+	+	N/A
Brake-controlled truing device	+	+	−	N/A
Mild steel-molybdenum block	0	−	−	N/A
Electroplated or metal-bond diamond block	+	+	0	N/A
Silicon carbide wheel	+	+	−	N/A

*Key:
+ Recommended
0 Possible under certain conditions
− Not recommended
†Electroplated CBN grinding wheels generally require no truing or conditioning before or during use.

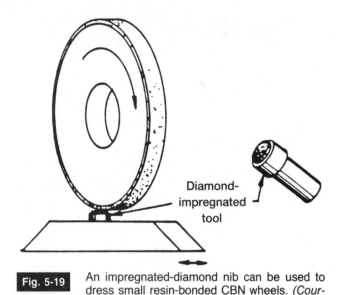

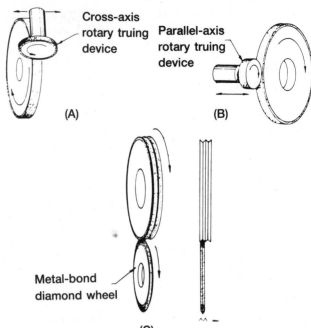

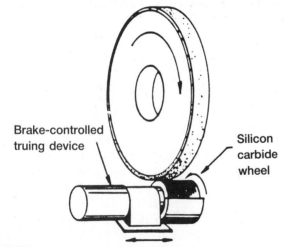

Fig. 5-19   An impregnated-diamond nib can be used to dress small resin-bonded CBN wheels. *(Courtesy of GE Superabrasives)*

Fig. 5-20   Brake-controlled truing devices use an abrasive grinding wheel to true CBN wheels. *(Courtesy of GE Superabrasives)*

Fig. 5-21   Rotary-powered truing devices use metal-bonded or electroplated diamond truing wheels for truing CBN wheels. *(Courtesy of GE Superabrasives)*

Diamond truing blocks (Fig. 5-22) can be either diamond-impregnated metal bond or electroplated with the abrasive in the 100- to 180-mesh size range. These truing blocks may be flat or have an intricate form to shape the grinding wheel to a desired shape or profile. Form block truing tools are most commonly used with resin-bonded CBN wheels; however they are also quite effective for truing some metal- and vitreous-bond wheels.

A metal dressing block (Fig. 5-23) made of soft steel or molybdenum is sometimes used for truing straight-faced CBN resin-bonded wheels. This is done by grinding the blocks with the untrued, undressed grinding wheel until it is worn into truth. The soft metal being ground will wear the CBN wheel quickly and will open the wheel face, exposing sharp crystals. Care must be used so that a wheel is not overtrued and its surface left with a very rough texture and many pockets where abrasive crystals have been dislodged from the wheel.

Nonabrasive crush rolls (Fig. 5-24) are used to crush-true grinding wheels made with special CBN bonds. These bonding systems allow the superabrasive wheel, which is run at low wheel speed, to be trued to the desired form and concentricity similar to the way conventional abrasive wheels are crush-trued. This truing method is an excellent system to generate and maintain intricate grinding-wheel forms.

DRESSING

The truing process usually leaves the working surface of the wheel smooth, with little or no abrasive crystal protrusion or clearance for chip generation and removal (Fig. 5-25). A wheel in this condition would burn the workpiece and remove little or no work material. *Dressing* is the process of removing some of the bond material from the surface of a trued wheel to expose the abrasive crystals and make the wheel grind efficiently.

A properly trued and dressed wheel will:

1. Produce workpieces of the required geometry, tolerance, and surface finish
2. Draw a minimum of grinding power
3. Produce workpieces without burn, surface damage, or chatter marks

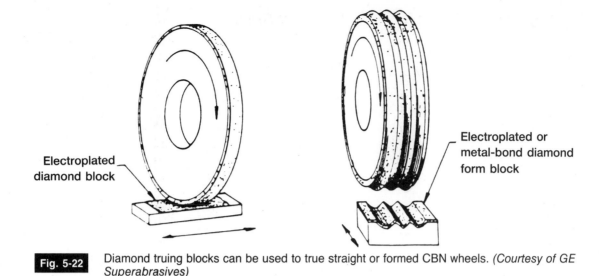

Fig. 5-22 Diamond truing blocks can be used to true straight or formed CBN wheels. *(Courtesy of GE Superabrasives)*

4. Increase the material removal rate and reduce grinding costs

Dressing Devices and Guidelines

A large number of dressing methods and equipment have been developed for CBN wheels over the years. A number of these methods or techniques have been built into some of the newer CBN grinders for automated and numerically controlled grinding systems to automate the dressing operation. The most common dressing methods or techniques are listed in Table 5-4 along with the types of CBN grinding-wheel bond system for which each is recommended.

An *aluminum oxide dressing stick or block* (Fig. 5-26A) is the simplest and most popular method of dressing or conditioning CBN grinding wheels. The operation consists of bringing a soft (G or similar grade), fine-grit (220-grit or finer) aluminum oxide stick into con-

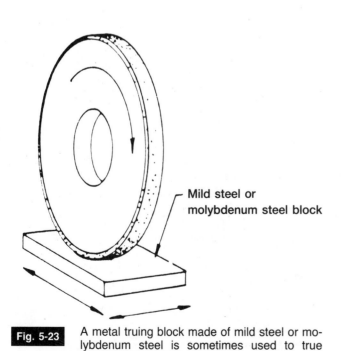

Fig. 5-23 A metal truing block made of mild steel or molybdenum steel is sometimes used to true straight-faced CBN resin wheels. *(Courtesy of GE Superabrasives)*

Fig. 5-24 Nonabrasive crush rolls are used to crush true CBN wheels made with a special bond. *(Courtesy of GE Superabrasives)*

Preparing the CBN Wheel and Grinder 57

After truing

The wheel face is smooth and closed

After dressing

The wheel face is open with the grits exposed, ready for efficient grinding action

After dressing

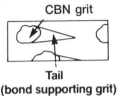
CBN grit

Tail
(bond supporting grit)

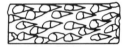
Path connecting the tails for coolant and chip flow

Fig. 5-25 Dressing opens the wheel face, exposing sharp abrasive crystals to make the wheel grind efficiently. *(Courtesy of Norton Co.)*

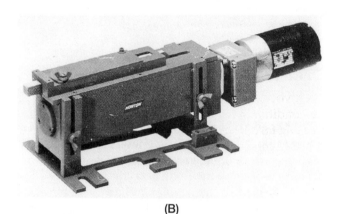

(A)

Fig. 5-26A An aluminum oxide dressing stick is the most popular method of dressing or conditioning a CBN grinding wheel. *(Courtesy of GE Super-abrasives)*

(B)

Fig. 5-26B Automated stick dressing devices are available to suit a variety of grinding machines. *(Courtesy of Norton Co.)*

tact with the rotating wheel surface. A small flow of coolant creates a slurry between the dressing stick and the revolving wheel which erodes the bond matrix and exposes sharp abrasive crystals. For safety reasons, it is advisable to mount the dressing stick in a vise or fixture and feed the wheel across the dressing stick. Automated stick dressing devices (Fig. 5-26B), which feed the stick into a revolving wheel at a constant rate, are available to suit a wide variety of grinders.

Water-soluble wax sticks (Fig. 5-27) consist of loose aluminum oxide grains mixed with polyethylene glycol, a water-soluble wax, and cast into the shape of a stick. In this dressing method, the grinding wheel must be positioned in close contact with the workpiece or similarly shaped metal part. As the wax stick is applied to the revolving wheel face, the abrasive grains carried by the wax are squeezed between the wheel and work, which acts as a pinch roll, eroding the bond matrix.

Another system, similar to the water-soluble wax stick because it also uses a pinch roll, has been incor-

Table 5-4 RECOMMENDED DRESSING AND CONDITIONING METHODS FOR CBN WHEEL BOND SYSTEMS

	Wheel Bonds*			
	Resin	**Metal**	**Vitreous**	**Electroplated[†]**
Aluminum oxide dressing stick	+	+	−	N/A
Loose abrasive	+	−	−	N/A
Soft steel-molybdenum grinding	+	−	−	N/A
Single-point diamond nib	0	−	0	N/A
Brake-controlled truing device	0	0	−	N/A
Wire brush	+	−	−	N/A
Rotary-powered diamond dresser	0	0	+	N/A

*Key:
+ Recommended
0 Possible under certain conditions
− Not recommended
[†]Electroplated CBN grinding wheels generally require no truing or conditioning before or during use.

porated on a number of special grinding machines designed to use only CBN wheels. This system feeds a mixture of aluminum oxide grain in a coolant between the roll and the grinding wheel by a special nozzle (Fig. 5-28). The wheel and the roll operate at nonsynchronous (different) speeds. This difference in speed creates a scuffing action between the wheel and the aluminum oxide grain which erodes the CBN wheel bond.

Stick Dressing Hints

STICK SELECTION

- Use conditioning sticks one or two grit sizes finer than CBN wheel grit. J-grade sticks are generally recommended.

- Use softer grade (H) abrasive sticks if less aggressive dressing action is required.

- For metal bonds, L grade is recommended.

STICK VOLUME TO BE USED.

The volume or length of the conditioning stick to be used is a function of the dressing force or infeed applied, wheel specifications, grinding conditions, and surface finish required. The following recommendations are given as good starting points:

- Use 0.05 to 0.10 in.^{3} of abrasive stick per square inch of the wheel face for *resin-bonded* wheels. *Example:* a 10-in.-diameter wheel, 1/2 in. thick, has 15.7 in.^{2} area. Using the above guidelines and 1/2 in. × 1/2 in. stick, it would require about 3 to 6 in. of stick length for adequate dressing.

Use the higher values for aggressive dressing action.

Use the lower values for mild dressing action.

Fine-grit wheels require less dressing than coarse grit wheels.

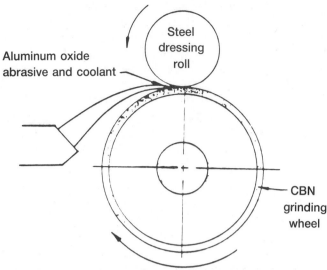

Fig. 5-28 A wheel being dressed by feeding a mixture of aluminum oxide abrasive and coolant between the dressing roll and the grinding wheel. *(Courtesy of GE Superabrasives)*

- Use 0.10 to 0.30 in.^{3} of conditioning stick per unit area of wheel face of the metal bond wheel.

- Vitrified bonded wheels generally do not require dressing. However, because of truing conditions or wheel specification chosen, it may occasionally be necessary to dress vitrified-bond superabrasive wheels. In case, such as an aid in crush truing, extremely light pressure and very small amount of stick volume should be used.

NOTE Most CBN wheel manufacturers include a dressing stick packed in the box with the CBN wheel. This is a helpful and constructive guide by the manufacturer. However, the size of the

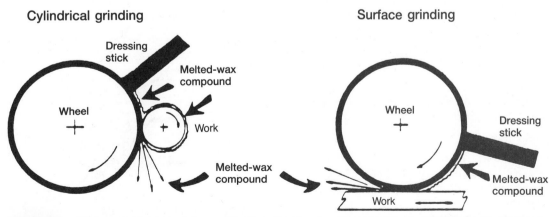

Fig. 5-27 Water-soluble wax sticks are ideally suited for dressing resin-bonded CBN wheels. *(Courtesy of GE Superabrasives)*

stick usually included will in most cases be insufficient to provide dressing for the life of the wheel. Never rely solely on the amount of stick in the box, but always have extra lengths of appropriate dressing stick available when using CBN wheels.

Loose aluminum oxide abrasive dressing methods have been developed to dress CBN wheels. The method illustrated in Fig. 5-29 uses an ejecting nozzle through which loose abrasive is blasted against the grinding wheel face by compressed air. The amount of "dress" can be controlled by the air pressure, grit size, and duration of the blast. This system has been automated and used on a grinder designed to finish grind hardened gears. The method illustrated in Fig. 5-30 uses a mixture of loose abrasive and coolant. An ultrasonic device drives the grains of aluminum oxide into the grinding wheel face eroding the bond matrix. Although this method is still experimental, it has proved very effective in test applications.

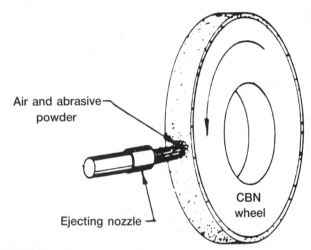

Air and abrasive powder

Ejecting nozzle

CBN wheel

Fig. 5-29 Dressing a CBN wheel by blasting aluminum oxide grains against the wheel face with a jet of high-pressure air. *(Courtesy of GE Super-abrasives)*

Wire brush dressing (Fig. 5-31) is another experimental dressing method which seems to show promise. The wires in the brush are stiff enough to erode the grinding wheel's bond matrix but flexible enough not to be ground away. The brush's axis of rotation is usually set at 45° to that of the grinding wheel's and can be self-powered or driven by the rotation of the grinding wheel.

Three other methods for dressing and conditioning a bonded CBN grinding wheel are more often described as truing devices.

1. A silicon carbide wheel on a brake-controlled truing device will leave a resin- or metal-bonding CBN wheel in a lightly dressed condition after the truing operation. The wheel can then be used to grind parts, usually at lower material removal rates, until it becomes fully dressed.

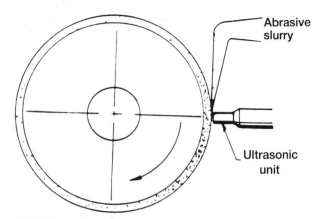

Abrasive slurry

Ultrasonic unit

Fig. 5-30 An ultrasonic dressing tool drives the grains of aluminum oxide into the wheel face to erode the bond material. *(Courtesy of GE Super-abrasives)*

2. Truing a CBN wheel by grinding a mild steel or molybdenum workpiece generally leaves the wheel in a partially dressed condition. When this wheel is used, it will condition itself fully during the grinding operation.

3. A single-point diamond nib can sometimes be used when dressing small vitreous-bonded CBN wheels. In most cases, these wheels have an open bond structure which allows the one-step truing and dressing method to be successful.

Conditioning

A properly specified CBN wheel will produce good-quality parts with a minimum of grinding power after correct truing and dressing. It may be necessary occasionally to condition the CBN wheel to restore the geometry of the part, restore the surface finish, or

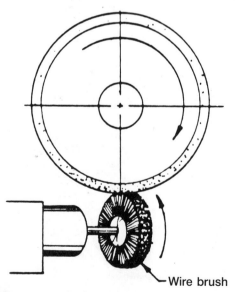

Wire brush

Fig. 5-31 A rotary wire brush is a method of dressing some CBN wheels. *(Courtesy of GE Super-abrasives)*

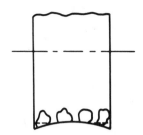

When wheel is out of form restore geometry by precision truing.

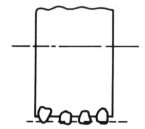

When bond erosion produces poor surface finish restore the wheel face by precision truing.

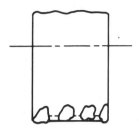

When CBN grit is worn out and grit exposure is reduced restore freeness of cut by precision dressing.

Fig. 5-32 Typical wheel situations that require conditioning. A correctly specified CBN wheel, when engineered properly, produces good quality parts at minimum of grinding power after correct truing and dressing. During its life, the CBN wheel will require periodic correction to its face, which is called *conditioning*. *(Courtesy of Norton Co.)*

both. Conditioning may involve a small amount of truing or dressing, or both (Fig. 5-32). Common dressing problems, their causes and their suggested remedies, are shown in Table 5-5.

TRUING AND DRESSING CBN WHEELS

It is not unusual for CBN wheels to outperform conventional grinding wheels by a margin of 250:1. Simple, but essential, truing, dressing, and coolant-application methods will ensure that cubic boron nitride grinding wheels live up to their reputation and justify their investment. The proper truing and dressing of the grinding wheel is probably one of the *most important factors* of all in making a CBN wheel work. An out-of-round wheel will pound the workpiece, reduce wheel life, and produce poor surface finishes. When properly conditioned, these wheels have the capability of increasing productivity, reducing overall grinding costs, consistently producing parts to close tolerances, and leaving the workpiece surface with no metallurgical damage.

Four major bond systems used in the manufacture of CBN wheels make up the largest percentage of wheels used today. These include resin bond, metal bond, vitreous bond, and electroplated bond. Each bond system is different in its requirements and responses to wheel conditioning procedures. *Impregnated bonds*, those where the abrasive is mixed with the bond and molded to form the abrasive section, generally require a two-step process (truing and dressing) to condition the wheel. This is particularly true with resin- or metal-bond systems and may be required with some types of vitreous bonds. *Electroplated grinding wheels* are the only type that do not require truing or dressing before use.

General Truing Guidelines

The following guidelines are offered to assist the operator in preparing a CBN wheel that will provide long and effective grinding performance.

1. Leave the CBN wheel and the wheel adapter together as a unit, if at all possible. This will prolong wheel life and save valuable retruing time.
2. Use a dial indicator and mount the wheel to within 0.001 in. (0.02 mm) *maximum runout* (Fig. 5-33).
3. When using large-diameter wheels or when operating at high wheel speeds, proper wheel balance and spindle bearing quality are very important.
4. It is advisable to true a CBN wheel on the machine on which it will be used.
5. Position the truing devices approximately ½ in. (13 mm) to the left of wheel center on a surface grinder; at or slightly below center on a cylindrical grinder (Fig. 5-34).
6. Use low wheel surface speeds, 1000 to 2000 sf/min (5 to 10 m/s) if possible. At a speed in this range, truing is fast and the wear rate on the truing device is generally lower.
7. Always use a large volume of grinding fluid to cool and lubricate the truing operation.
8. The cross travel of the truing device should be a smooth, steady motion with no hesitation.
9. Do not true the wheel more than is necessary. Excessive truing wastes expensive wheel abrasive and shortens wheel life.

Truing and Dressing Procedures

Since there is a variety of truing devices, some of which apply to only one bond system while others apply to various bond systems, the truing procedure for each will be covered in detail.

DIAMOND-IMPREGNATED NIB

Used for truing 8 in. (200 mm) and smaller wheels:

Resin-bond wheels
Some vitreous-bond wheels

Table 5-5 DRESSING PROBLEMS, CAUSES, AND REMEDIES

Problem	Cause	Remedy
1. Wheel causes burn	Wheel face not sufficiently open Improper dressing or improper grinding	Check for grit exposure Check coolant flow—is it directed at the grinding zone? Use more stick length, additional dressing stick
2. Wheel cannot be dressed open	Insufficient dressing Wheel severely glazed or closed from severe truing	Apply progressively higher dressing pressure Use coarser grit, softer-grade stick, avoid harsh truing conditions
3. Poor surface finish	Severe dressing Overdressing	Use softer or finer grit stick Lower dressing force Decrease abrasive volume used for dressing
4. Burnished work surface finish	Excess friction and rubbing Wheel face not sufficiently open	Severe truing conditions leading to flats on superabrasive grits Use softer grade stick, increase dressing forces Increase abrasive stick volume used
5. Wheel is loaded with chips	Improper dressing Wheel face not sufficiently open Severe grinding condition	Increase coolant flow and pressure to wash off the chips Increase volume of dressing stick and dressing force used Reduce the material removal rate Check wheel for proper specification (grit size, concentration, bond type, and grade)

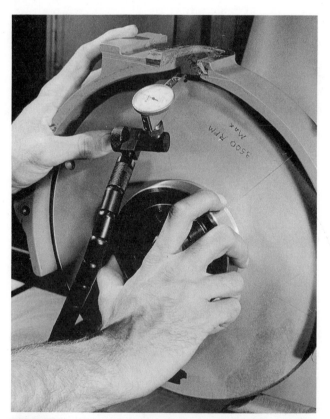

Fig. 5-33 Use a dial indicator to mount the CBN wheel to within 0.001-in. (0.02-mm) maximum runout. *(Courtesy of GE Superabrasives)*

Procedure

1. Mount the wheel on the grinder spindle and indicate it to within 0.001-in. (0.02-mm) or less maximum runout.
2. Tighten the wheel securely on the grinder spindle.
3. Clean the magnetic chuck thoroughly with a cloth.

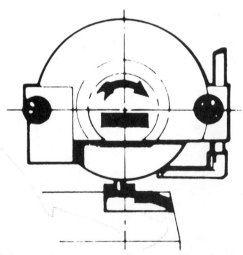

Fig. 5-34 Position the truing device approximately ½ in. (13 mm) to the left of the wheel center on a surface grinder. *(Courtesy of GE Superabrasives)*

4. Place a piece of smooth paper, no more than 0.002 in. (0.05 mm) thick and slightly larger than the base of the diamond holder, on the left-hand end of the magnetic chuck.

NOTE The paper prevents scratching, marring, or destroying the accuracy of the magnetic chuck surface when the holder is removed.

5. Place the diamond holder covering as many magnetic inserts as possible and energize the chuck (Fig. 5-35).
6. Raise the CBN wheel above the height of the diamond nib.
7. Move the grinder table longitudinally so that the diamond nib is about ½ in. (13 mm) to the left of the wheel centerline (Fig. 5-35).

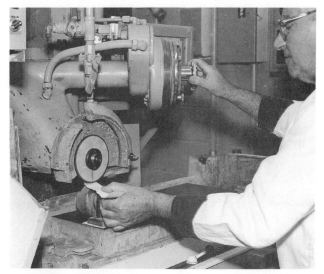

Fig. 5-36 With a piece of paper between the diamond nib and the wheel, lower the wheelhead until contact is made with the paper. *(Courtesy of M. Rapisarda and GE Superabrasives)*

Fig. 5-35 Secure the diamond holder on the left end of the grinder table. *(Courtesy of M. Rapisarda and GE Superabrasives)*

8. Lock the grinder table in this position.
9. Hold a piece of paper between the wheel and the diamond nib and lower the wheelhead until it just touches the paper (Fig. 5-36).
10. Revolve the wheel by hand to be sure that the wheel does not contact the diamond truing nib. If contact is made with the nib, raise the wheel about 0.003 in. (0.07 mm).
11. Set the grinder spindle to run at 1000 to 2000 sf/min (5 to 10 m/sec) if possible.
12. Lightly coat the CBN wheel surface with a wax marking crayon (Fig. 5-37).
13. Feed the grinder wheelhead down in increments of 0.0004 in. (0.01 mm) or less until you hear the wheel contacting the diamond nib.
14. Start the grinder coolant to ensure that a good

supply is carried directly into the wheel-truing tool interface.
15. Feed the diamond across the wheel face at a rate of 3 to 12 in./min. Be sure that the wheel clears the diamond in both directions during the truing operation. *Never* allow the truing tool to dwell under the wheel.
16. Continue truing passes of increments of 0.0004 in. (0.01 mm) or less until the crayon is removed from the entire wheel circumference. *Do*

Fig. 5-37 Lightly coat the CBN wheel face with a wax marking crayon. *(Courtesy of M. Rapisarda and GE Superabrasives)*

not overtrue. It is wise to check the wheel occasionally to see if it is true; some of the crayon could be in the wheel voids. After a wheel has been trued, it will have a smooth surface (Fig. 5-38) and must be dressed to expose the sharp abrasive crystals.

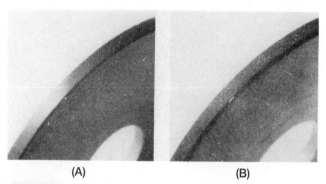

(A)	(B)

Fig. 5-38 A CBN wheel face. (A) After truing. (B) After dressing. *(Courtesy of GE Superabrasives)*

17. Remove the truing device from the grinder.
18. Select a 200-grit, G-grade aluminum oxide dressing stick, or appropriate grade for each specific wheel.
19. Saturate the dressing stick with coolant so that a slurry is created when it contacts the wheel.
20. Hold or clamp the dressing stick securely and bring it into contact with the revolving wheel (Fig. 5-39).

Fig. 5-39 Feed the dressing stick aggressively into the revolving wheel. *(Courtesy of M. Rapisarda and GE Superabrasives)*

21. Feed the dressing stick aggressively into the wheel to open the face of the wheel.

22. Continue dressing until the dressing stick wears quickly. This is a good sign that the wheel face is open and ready for grinding.

BRAKE-CONTROLLED TRUING DEVICES

Used for truing:

> Resin-bond wheels
> Metal-bond wheels

A brake-controlled truing device, generally using a 60-grit, L-hardness silicon carbide wheel, can be used to true resin and metal-bond CBN wheels. Although the truing procedures with a brake-controlled truing device are very similar to those used with a diamond-impregnated nib, there are some major differences.

Procedure

1. Mount the CBN wheel on the grinder spindle and indicate it to within 0.001 in. (0.02 mm) or less maximum runout (Fig. 5-33).
2. Tighten the wheel securely on the spindle.
3. Clean the magnetic chuck with a cloth.
4. Place a piece of smooth paper, no more than 0.002 in. (0.05 mm) thick and slightly larger than the base of the truing device, on the left end of the magnetic chuck.
5. Place the brake-controlled truing device so that its spindle is in line with the grinder spindle and energize the chuck (Fig. 5-40).
6. Raise the CBN wheel slightly above the height of the truing wheel.
7. Move the surface grinder table longitudinally so that the center of the truing wheel is about ½ in. (13 mm) to the left of the CBN wheel centerline (Fig. 5-40).
8. Lock the grinder table in this position.
9. Hold a piece of paper between the truing wheel and the stationary CBN wheel and lower the wheelhead until it just touches the paper (Fig. 5-41).
10. Revolve the CBN wheel by hand to ensure that it does not contact the truing wheel. If contact is made, the wheelhead should be raised about 0.003 in. (0.07 mm).
11. Set the grinder spindle to run at 1000 to 2000 sf/min (5 to 10 m/s) if possible.
12. Lightly coat the CBN wheel face with a wax marking crayon (Fig. 5-37).
13. Spin the truing wheel with a long pencil and feed the grinder wheelhead down in 0.001-in. (0.02-mm) increments until the two wheels are in contact.
14. Turn the grinder crossfeed handle so that the

Fig. 5-40 The center of the brake-controlled truing wheel should be located about ½ in. (13 mm) to the left of the center of the CBN wheel. *(Courtesy of M. Rapisarda and GE Superabrasives)*

truing wheel just clears the edge of the CBN wheel.

15. Adjust the grinder coolant so that a good supply is carried directly into the point of contact of the two wheels.

16. True the CBN wheel until the crayon is removed from the entire wheel circumference. The following truing downfeeds are recommended:

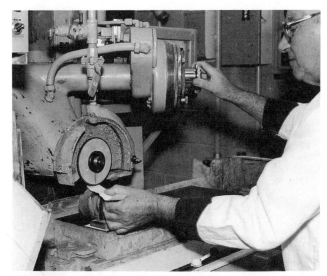

Fig. 5-41 Hold a piece of paper between the two wheels and lower the grinder wheelhead until they just touch the paper. *(Courtesy of M. Rapisarda and GE Superabrasives)*

a. Coarse-mesh wheels—0.001 to 0.002 in. (0.02 to 0.05 mm)
b. Fine-mesh wheels—0.0002 to 0.0005 in. (0.005 to 0.013 mm)
c. Finish passes—0.0002 in. (0.005 mm)

NOTE Reduce the downfeed rate if chatter occurs during the truing operation.

17. The traverse rate of the truing wheel should be between 3 and 12 in. (75 and 300 mm) per minute, and the truing wheel must clear the edge of the CBN wheel on each pass.

18. Inspect the CBN wheel occasionally to check whether it is true. *Do not overtrue.*

A smooth sound during the truing operation generally indicates that the wheel is true.

19. Remove the truing device from the grinder.

20. Select a 200-grit, G-grade aluminum oxide dressing stick, or the appropriate stick for the specific wheel.

21. Dip the dressing stick in coolant so that a slurry is created when it contacts the CBN wheel.

22. Hold the dressing stick securely and bring it into contact with the revolving wheel (Fig. 5-42).

23. Feed the dressing stick aggressively into the wheel to open the face of the wheel.

24. A good sign that the CBN wheel is dressed is when the dressing stick starts to wear away rapidly.

ROTARY-POWERED TRUING DEVICES

Used for truing wheels 8 in. or larger, and production grinding:

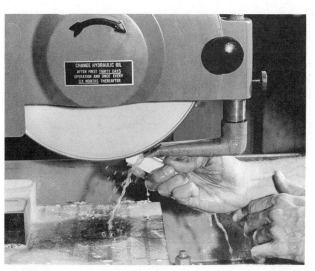

Fig. 5-42 An abrasive dressing stick being used to open the face of a wheel which has been trued. *(Courtesy of GE Superabrasives)*

Resin-bond wheels
Metal-bond wheels
Vitreous-bond wheels

Impregnated-bond and electroplated diamond wheels or rolls (Fig. 5-43) are recommended in rotary-powered truing devices for truing CBN wheels of 8 in. (200 mm) diameter or larger. Diamond-wheel truing systems are also effective on smaller CBN wheels of all types. The diamond abrasive in the truing wheel should be in the ASTM E-11 100/180-mesh (FEPA D252–D76) range. Use of larger diamond abrasive mesh size results in faster truing cycles; smaller mesh size produces a smoother, more accurately trued grinding surface.

Truing Device Hints

1. The rotational direction of the cutter can be set so that its velocity is in the same direction with the wheel at the point of contact (+) (unidirectional) or

Fig. 5-43 Types of rotary diamond cutters used for CBN wheel truing. *(Courtesy of Norton Co.)*

opposite at the point of contact (−) (counterdirectional), by proper arrangement of the truing device and the power input.
   a. In general, (+) rotational direction produces lower truing force and the best truing results.
   b. Mounting the cutter with its axis parallel to the wheel axis is preferred. However, if there are space constraints such as in I.D. grinders, then such orientation can be achieved using a cup-type cutter and the cutter axis normal to the wheel axis.
2. The ratio of cutter speed (in feet per minute) to wheel speed (in feet per minute) is called the *speed ratio* (SR). A speed ratio of +0.5 is a good starting point. Higher speed ratios up to +0.8 may be beneficial in some instances. Speed ratios as low as +0.2 can be used but will result in higher truing forces. Where possible, the SR can be increased by using large-size truing cutters. *To avoid damage to the device, do not exceed the rated speed of the truing device.*
3. Mount the wheel as close to truth as practical before attempting to true and dress. This saves time, money, and valuable CBN or diamond abrasive.
4. Whenever possible, balance the wheel properly before attempting to true and dress. For precise application with large diameter wheels, balance should be checked after initial truing as well.
5. The recommended truing infeed on radius per traverse is 0.0001 in. (the maximum infeed on radius per traverse is 0.0002 in.) for traverse truing.
6. In most cases, a truing lead of 0.002 to 0.006 in. per revolution is satisfactory. This will generally require a traverse rate in the range of 5 to 20 in./min.
   a. To achieve the recommended lead in I.D. grinders, higher traverse rates may be required. The truing cutter thickness/truing lead ratio is defined as the *overlap ratio*. Small overlap ratios (30 or less) generally render the wheel face more open and free cutting.
   b. Consideration of overlap ratio is very critical for vitrified-bond superabrasive wheels.
7. Plunge truing will require infeed increments of 1 to 5 millionths of an inch per revolution. Minimize the dwell of the cutter on the wheel at the end of truing to prevent dull wheel face.
8. Rotary truing with proper truing devices and the recommended truing conditions will minimize truing forces. However, if the wheel setup is not correct or excessive truing is required, then the truing process may lead to higher forces due to the dulling or glazing of the wheel face. In such cases (particularly for resin- or metal-bond wheels), intermittent stick dressing of the superabrasive wheel will help to maintain low truing forces. This prevents chatter and increases the ease of truing.

Fig. 5-44 A rotary-powered truing device using a diamond rotary cutter may be used for truing CBN wheels. *(Courtesy of Norton Co.)*

Truing Dos and Don'ts

1. Occasionally the diamond cutter may need to be stick-dressed, particularly in I.D. grinding operations.
   a. Always true the CBN wheels using diamond cutters with good coolant application, i.e., flow rate, pressure, and direction and position of coolant jet to the wheel face.
   b. Use full coolant flow available in the grinder.
2. The truing device is a precision unit. Mounting, use, and care for this device should be given proper attention to achieve the full benefit of this device.
3. *Do not run the truing device continuously. The device must be turned off after truing and during the grinding cycle to avoid overheating and damage to the device.*
4. If the rotational direction is being changed for the truing device, *be sure that the cutter comes to a complete stop before changing the direction,* for safety and to avoid damage to the device.
5. *Do not exceed the recommended speeds* for the truing devices.
6. Make sure that the devices are properly guarded during operation.

Procedure

1. Mount the CBN wheel on the grinder spindle and use an indicator to true it within 0.001 in. (0.02 mm) or less maximum runout.
2. Tighten the wheel securely on the grinder spindle.
3. Select the proper diamond truing wheel to suit the CBN wheel to be trued.
4. Fasten the truing wheel securely to the rotary-powered truing device.
5. Set the rotary-powered truing device on the grinder, being sure that the truing device and grinding wheel spindles are at the same height.
6. Set the truing wheel at any angle (30, 45, 60, or 90°) or parallel to the grinding-wheel axis (Fig. 5-44).
7. Fasten the truing device securely.
8. Locate the dressing wheel with the CBN wheel face.
9. Hold a piece of paper between the truing wheel and the stationary CBN wheel and bring the two wheels together until they just touch the paper.
10. Revolve the CBN wheel by hand to see that it does not contact the truing wheel.
    • If contact is made, adjust so that the wheels are approximately 0.003 in. (0.07 mm) apart.
11. Set the grinder spindle to run at 1000 to 2000 sf/min (5 to 10 m/s) if possible.
12. Lightly coat the CBN wheel face with a wax marking crayon (Fig. 5-37).
13. Adjust the grinder until the CBN wheel lightly contacts the truing wheel.
14. Traverse the grinder table so that the truing wheel clears the edge of the CBN wheel.
15. Adjust the grinder coolant so that a good supply is carried directly into the point of contact between the two wheels (Fig. 5-45).
16. Traverse the truing wheel across the face of the CBN wheel in an absolutely smooth motion; *there should be no hesitation.*
17. Continue the truing operation using infeeds of 0.0001 in. (0.002 mm) and be sure to clear the edge of the wheel on each traverse pass.
18. Continue truing passes in 0.0001-in. (0.002-mm) increments until the crayon is removed from the entire wheel circumference.
    Do not overtrue. It is wise to check the wheel occa-

Use a good supply of grinding coolant during the wheel truing operation. *(Courtesy of Norton Co.)*

sionally to determine whether it is true; some of the crayon may be in the wheel voids.

19. Use an aluminum oxide stick to dress the wheel to expose sharp abrasive grains. Common truing problems, their causes and their suggested remedies, are shown in Table 5-6.

NOTE When mixed properly, CBN wheels made with porous vitreous bonds do not usually require a separate conditioning operation. Because of their porous microstructure, they are quite open and free-cutting after the truing operation.

DIAMOND TRUING BLOCKS

Used for truing:

> Resin-bond wheels
> Metal-bond wheels
> Vitreous-bond wheels

Diamond-impregnated metal bond or electroplated truing blocks generally perform well for truing resin-bond CBN grinding wheels. They also have limited applications for vitreous- and metal-bond CBN grinding wheels. The truing surface of these blocks may be flat or formed, depending on the

wheel shape required. Some manufacturers who produce form grinding wheels also provide a metal-bonded diamond forming block or tool of the same form as a matched set (Fig. 5-46).

Procedure

1. Mount the CBN wheel on the grinder and indicate within 0.001 in. (0.02 mm) maximum runout.
2. Select a diamond truing block to suit the shape of the CBN wheel to be trued. The diamond abrasive mesh size should be in the ASTM E-11 100/ 180 (FEPA D252–D76) range.
3. Locate the truing block on the grinder table and energize the magnetic chuck.
4. Hold a piece of paper between the CBN wheel and the top of the truing block and lower the wheel until it just touches the paper.
5. Revolve the CBN wheel by hand to see that it does not contact the truing block. If contact is made, adjust the wheel so that it is about 0.003 in. (0.07 mm) above the truing block.
6. Set the grinder spindle to revolve at 1000 to 2000 sf/min (5 to 10 m/s) if possible.
7. Lightly coat the CBN wheel face with a wax marking crayon.
8. Bring the grinding wheel down until it just lightly contacts the top of the truing block.
9. Traverse the grinder table so that the CBN wheel clears the truing block.
10. Adjust the grinder coolant so that a good supply

Some CBN form wheel manufacturers supply a form truing block to match the wheel shape or contour. *(Courtesy of GE Superabrasives)*

Table 5-6 TRUING PROBLEMS, CAUSES, AND REMEDIES

Problem	Cause	Remedy
1. Wheel runout increases and is not decreased by truing	Excess truing force causing deflection	Lower truing forces Smaller infeed Intermittent wheel dressing
2. Smell of bond burning (resin bond); ringing noise during truing	Severe truing conditions	Lower frictional heat by intermittent dressing Lower truing forces by reducing infeed or traverse rate Increase coolant flow and pressure Use finer grit diamond rotary cutter (proper truing should generate barely audible noise)
3. Wheel cannot be trued to within recommended maximum of 0.0002 in.	Lack of system rigidity and/or severe truing conditions	Check consistency of infeed Decrease the infeed Gentle intermittent dress using soft conditioning stick Check machine spindle looseness or runout Check for runout of the diamond rotary cutter Assure rigid mounting
4. Truing time is too long	Excess truing or improper procedure	Check wheel mounting procedure (wheel should be within 0.002 in. when initially mounted) Avoid severe truing
5. Wheel surface shows excessive grit pullout.	Excess truing force	Check the truing infeed—is it too high? Avoid severe truing conditions
6. Excess diamond rotary cutter wear or taper on workpiece	Improper truing	Check for proper diamond rotary cutter selection Check for proper truing procedures Check for proper traverse rate

is carried directly into the wheel-truing block interface.

11. Traverse the CBN wheel across the full length of the truing block in a smooth, steady motion.
12. Continue the truing operation using infeeds of 0.0002 in. (0.005 mm) per pass. Several passes may be required before the wheel is true.
13. Be sure to clear the edge of the truing block on each traverse pass. Examine the wheel occasionally to see if it is true.
14. Dress the wheel face with a dressing stick to expose sharp abrasive grains.

ELECTROPLATED WHEELS

Electroplated CBN grinding wheels are the only type that do not require truing or dressing before use. These wheels are made to run true by careful mounting. An electroplated CBN wheel has the proper roundness, concentricity, and profile built in during manufacture. Most electroplated wheels have an indicating ring machined into the wheel core (Fig. 5-47). With the aid of an indicator, align the wheel within 0.0002 in. (0.005 mm) or less maximum runout.

CRUSH-TRUING SYSTEMS

Several manufacturers of CBN wheels now offer wheels with crushable bonds. These wheels can be trued to the desired concentricity and form by using hardened-steel or carbide crush rolls, similar to those used to crush-true aluminum oxide wheels (Fig. 5-48). Intricate grinding-wheel forms can easily be generated and maintained with these rolls. Always follow wheel manufacturers' guidelines for crush-truing.

TRUING PROCEDURES ON VARIOUS MACHINES

Since CBN wheel conditioning methods may vary somewhat according to the application or grinder on which they are used, the following guidelines are offered to assist the machine-tool operator. Keep in mind that these are only general suggestions and it is good practice to follow the wheel manufacturer's instructions for the truing and dressing procedures for various types of CBN wheels.

Surface Grinder

- Use a diamond-impregnated tool, diamond truing block, or brake-controlled truing device with a silicon carbide wheel (Fig. 5-49).

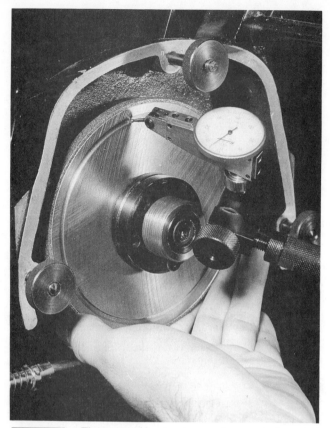

Electroplated wheels require no truing and dressing but must be mounted within 0.0002-in. (0.005-mm) maximum runout. *(Courtesy of GE Superabrasives)*

- Run the CBN wheel at normal grinding speed or if possible, at 1000 to 2000 sf/min (5 to 10 m/s).

- Use downfeed rates of 0.0002 in. (0.005 mm).

- Use a good supply of coolant during truing.

- Use a traverse rate of 3 to 12 in. (75 to 300 mm) per minute.

Fig. 5-48   Crush truing a CBN vitreous wheel manufactured with a crushable bond.

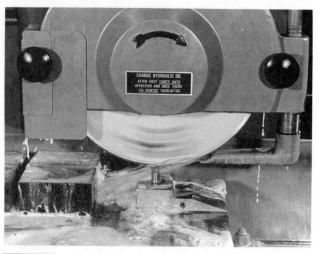

Fig. 5-49   A good supply of coolant prevents damage to the truing device while conditioning a wheel on a surface grinder. *(Courtesy of GE Superabrasives)*

- Traverse off the wheel in both directions.

- Dress the wheel.

Cylindrical Grinder

- Use an 80- to 150-grit diamond-impregnated tool or rotary truing device (Fig. 5-50).

- Mount the truing device on the machine table.

- Use truing infeeds of 0.0002 in. (0.005 mm) or less, depending on the truing device used.

- Use a good supply of coolant during the truing operation.

- Traverse the table at a rate of 12 in. (300 mm) per minute to move the truing tool back and forth over the wheel face.

- Remember to dress the wheel.

Internal Grinder

- Use a rotary-powered diamond-impregnated wheel or a diamond-impregnated tool (Fig. 5-51).

- Traverse at the rate of 15 to 25 ft (375 to 625 mm) per minute.

- Use truing infeeds of about 0.001 in. (0.02 mm).

- Use a good supply of coolant during the truing operation.

- Remember to dress the wheel, if necessary.

Tool and Cutter Grinder

- A Type II V9 cup wheel is normally used which generally needs no truing (Fig. 5-52).

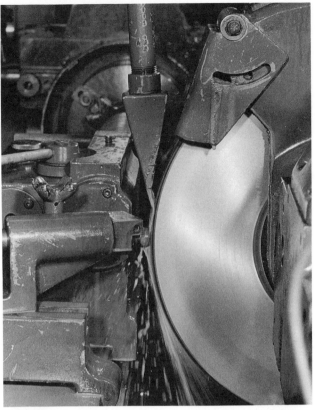

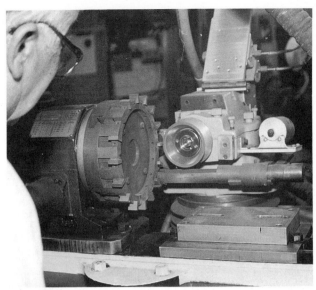

Fig. 5-52 A Type IIV9 CBN cup wheel is generally used on tool and cutter grinders. *(Courtesy of M. Rapisarda and GE Superabrasives)*

- If a runout check shows that truing is required, use a diamond-impregnated tool or a brake-controlled device.

- Use infeed passes of 0.001 in. (0.02 mm) or less.

- Use coolant, if possible.

- Don't forget to dress the wheel.

Fig. 5-50 Truing a wheel on a cylindrical grinder with a diamond-impregnated tool. *(Courtesy of GE Superabrasives)*

Fig. 5-51 An internal grinding wheel being trued with a rotary-powered diamond-impregnated wheel. *(Courtesy of Norton Co.)*

Fig. 5-53 Truing a CBN jig grinder wheel with a rigidly mounted diamond-impregnated nib. *(Courtesy of GE Superabrasives)*

Jig Grinder

- Use a diamond-impregnated nib with 150-mesh abrasive for most wheel truing (Fig. 5-53). Use a single-point nib to true resin- and vitreous-bond wheels if the diamond point is sharp.

- Use infeed passes of 0.0002 in. (0.005 mm).

- Apply a spray mist of water-soluble oil during the truing operation, if possible.

- Use fast traverse rates (30 to 40 in. or 750 to 1000 mm per minute).

- *Dress the wheel;* vitreous-bond wheels may not require dressing.

Centerless Grinder

- Use a power-driven rotary diamond truing roll.

- Set the truing roll at cross axis or parallel to the wheel axis (Fig. 5-54).

- Use infeed rates of 0.0002 in. (0.005 mm).

- Apply a good supply of coolant while truing.

- Use a traverse rate of 6 to 12 in. (150 to 300 mm) per minute.

- Dress the wheel, if required.

Production Grinders

Some of the new grinders produced specially to use CBN grinding wheels use electroplated wheels which

Fig. 5-54 A parallel-axis rotary diamond truing device being used to true a centerless grinder wheel. *(Courtesy of GE Superabrasives)*

require no truing or dressing. Other grinders use preconditioned wheels supplied by the manufacturer and require no conditioning for the life of the wheel.

For other grinders using metal- or resin-bond wheels, new, innovative, and automatic truing and conditioning systems are available. Figure 5-55 illustrates a typical production truing and conditioning system where the operations can be pushbutton- or computer-controlled and do not physically require the operator.

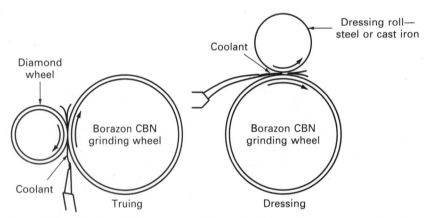

Fig. 5-55 A typical production truing and conditioning system which can be automated or operator-controlled. *(Courtesy of GE Superabrasives)*

REVIEW QUESTIONS

Grinding Machines

1. List five characteristics which a grinder should have to use CBN wheels effectively.
2. Why is it important that a grinder be able to maintain constant speeds when using CBN grinding wheels?
3. Why is it so important to properly true and dress a CBN wheel?

CBN Grinders

4. Name the five characteristics of production grinding machines specifically designed for CBN wheels.

Mounting the Wheel

5. Why is it important that a CBN wheel be mounted properly on a grinder spindle?
6. Explain why it is advisable to leave the CBN wheel and the adapter together as a unit for the life of the wheel.
7. What is the purpose of placing a mark on the wheel adapter and the grinding wheel?
8. How close should a wheel be trued with an indicator?
9. What type of fit should there be between the CBN wheel and the grinder spindle or adaptor?

Truing

10. State three purposes for truing a wheel.
11. Name two types of diamond nibs which may be used for truing CBN wheels.
12. Briefly describe a brake-controlled truing device and list the CBN wheels for which they can be used.
13. For what purpose are rotary-powered truing devices used?

Dressing

14. State two purposes of dressing a CBN grinding wheel.
15. Name two types of abrasive sticks used for dressing CBN wheels.
16. Briefly explain how dressing sticks help to open a CBN wheel.
17. How is the amount of "dress" controlled when using loose aluminum oxide dressing methods?

Truing and Dressing CBN Wheels

18. What is the most important factor in making a CBN wheel grind effectively?
19. How will an out-of-round CBN wheel react?
20. What type of CBN wheels require no truing and dressing?
21. Explain the following general truing guidelines:
    a. wheel runout
    b. position of truing device
    c. wheel speed

Diamond-Impregnated Nib

22. For what purpose are diamond-impregnated truing nibs used?
23. How can the CBN wheel be accurately set to the diamond nib?
24. How can you tell when a CBN wheel is trued?

Brake-Controlled Truing Devices

25. What type of wheel is commonly used on brake-controlled truing devices?
26. How should the brake-controlled truing device be located on a surface grinder?
27. Describe the dressing stick which should be used to dress resin- and metal-bond CBN wheels.

Rotary-Powered Truing Devices

28. What type of wheels or rolls are generally used with rotary-powered truing devices?
29. Explain how a rotary-powered truing device should be set on a grinder.
30. What is the recommended infeed when using rotary-powered truing devices?

Diamond Truing Blocks

31. What type of wheels can be trued with diamond-impregnated truing blocks?
32. Name two types of truing blocks.

Truing Procedures on Various Machines

33. Describe an automatic truing and dressing system available on new CBN grinders.

CHAPTER SIX

Economics of Superabrasive Tools

The productivity gap between the United States and overseas workers and the intense global competition are forcing many companies toward a renewed commitment to produce quality products while at the same time to reduce manufacturing costs. An understanding of the productivity capabilities of superabrasive grinding wheels and polycrystalline cutting tools is necessary in order to appreciate the benefits that superabrasive tools offer metalworking industries. Many people consider the cost of these *high-efficiency* cutting tools to be high; actually, their cost is very low when factors such as long tool life, increased productivity, and better-quality products are considered.

OBJECTIVES

After completing this chapter you should be able to:

1. Understand each element of a superabrasive tool cost
2. Understand the difference between the *cost of the tool* and the *cost of using the tool*
3. Perform a cost analysis (justification) for superabrasive grinding wheels and polycrystalline blank tools

ROADBLOCKS TO JUSTIFICATION

In most companies, when new machine tools or superefficient cutting tools are being considered for purchase, the cost accounting department must determine whether the equipment can be justified. New superabrasive tools can cost from 10 to 50 times more than conventional tools, and this apparent large increase in tool costs seems to present problems for the cost accounting department. Under the previous costing system, even though the new equipment should be an improvement over present methods, the numbers generally say that the company cannot afford the expenditure.

Under the previous costing system, several common errors are committed in the area of the cost-benefit assumptions. The most common of these costing errors are:

1. The overemphasis of direct labor reduction.
2. Failure to recognize the importance of product quality.
3. Failure to recognize the advantages of a new cutting tool or new machine tool.

Direct Labor Reduction

Management frequently overemphasizes the importance of direct labor reduction. In the past, direct labor often constituted as much as 50 percent of the total product cost, and traditional return-on-investment (ROI) analysis could justify new equipment or tools on the basis of direct labor savings. Today, direct labor only accounts for about 10 percent of the total manufacturing costs. Material costs now make up 55 percent of the total product cost, overhead about 20 percent, and indirect labor about 15 percent (Fig. 6-1). Savings in material costs, overhead, and indirect labor are areas which are often more important criteria for justification than are direct labor savings.

Importance of Quality Products

A common pitfall in justifying expenditures on cutting tools and equipment is the failure to consider the positive effect that an improvement in product quality can have on the productivity of an entire company. Many expenditures can be justified on reduced warranty claims, scrap, and rework costs that are a

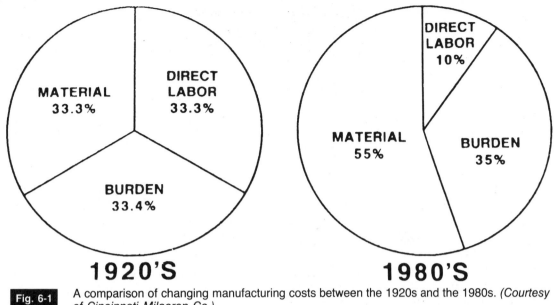

MATERIAL 33.3%   DIRECT LABOR 33.3%   BURDEN 33.4%

1920'S

DIRECT LABOR 10%   MATERIAL 55%   BURDEN 35%

1980'S

Fig. 6-1 A comparison of changing manufacturing costs between the 1920s and the 1980s. *(Courtesy of Cincinnati Milacron Co.)*

direct result of an improvement in the manufacturing process. Even a small improvement at every workstation would have a dramatic effect on reducing manufacturing costs. Most companies succeed because of their reputation of producing quality products, while those producing poor-quality products quickly lose customers and eventually go out of business.

NEW TOOL ADVANTAGES

To obtain the lowest possible costs in any grinding or machining operation, tool engineers and methods personnel must look at the entire operation and examine each factor which contributes to the cost picture.

A new high-efficiency cutting tool or a new machine tool offers many advantages, and the effect of each on the manufacturing process must be considered. Since most of the material cut in the metalworking industry is ferrous metal, this book will deal primarily with the use of cubic boron nitride (CBN) grinding wheels and PCBN tool blanks. The advantages of each tool (grinding wheel or tool blank) are very similar, and Fig. 6-2 shows the advantages which apply primarily to PCBN tools.

Long Tool Life. Superabrasive CBN cutting tools have properties which resist chipping and cracking and provide uniform hardness and abrasion resistance in all directions. They may outperform conventional cutting tools by 20 to 100 times. Reduced tool wear results in closer tolerances on workpieces, and fewer tool adjustments keep machine downtime to a minimum.

High Material Removal Rates. Because these cutting tools resist abrasion so well, cutting speeds in

the range of 250 to 6000 sf/min (274 m/min) and feed rates of 0.010 to 0.020 in. (0.25 to 0.50 mm) are possible. This results in higher material-removal rates with less tool wear, which reduces the total machining cost per piece.

Cutting Hard, Tough Materials. These cutting tools are capable of efficiently machining all ferrous materials with a hardness of Rc 45 and above and for machining cobalt-base and nickel-base high-temperature alloys with a hardness of Rc 35 and above. Many of these materials were so hard and abrasive that grinding, which is a relatively slow metal-removal process, was previously the only practical way to machine them.

High Quality Products. Because the cutting edges of PCBN cutting tools wear very slowly, they produce high-quality parts faster and at a lower cost per piece than do conventional cutting tools. The need for the inspection of parts is greatly reduced, as is the adjustment of the machine tool to compensate for cutting tool wear or maintenance. This results in better control over workpiece shape and size, thus producing consistent part quality far beyond that possible with other cutting tools.

Uniform Surface Finish. Surface finishes in the range of 20 to 30 microinches (μin.) are possible during roughing operations with PCBN cutting tools. On finishing operations, surface finishes in single-digit microinch numbers are possible, which can eliminate the need for relatively slow grinding operations, which, in turn, reduces total manufacturing costs.

Lower Cost per Piece. PCBN cutting tools remain sharp and cut efficiently through long production

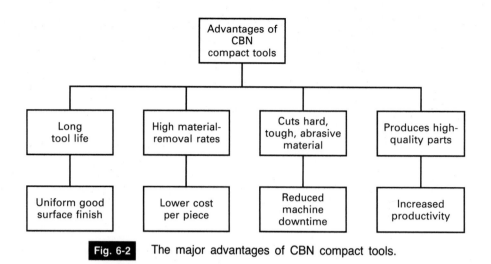

Fig. 6-2 The major advantages of CBN compact tools.

runs. This results in consistently smoother surface finishes, better control over workpiece shape and size, and fewer cutting-tool changes. This also lowers manufacturing costs per piece by reducing inspection time and increasing machine uptime.

Reduced Machine Downtime. Since PCBN cutting tools remain sharp much longer than do carbide or ceramic cutting tools, there is less time required to index, change, or recondition the cutting tool. This results in less machine downtime, which increases the amount of time that value is added to in-process inventory and, therefore, reduces the total manufacturing cost per piece.

Increased Productivity. A combination of all the advantages that PCBN cutting tools offer, such as increased speeds and feeds, long tool life, longer production runs, consistent part quality, and savings in labor costs, all have an effect on the overall manufacturing cost per piece. The positive effects that improved quality has on the productivity of an entire company cannot be overlooked. A company's reputation can be enhanced by producing quality products at competitive prices, while poor-quality products adversely affect a company's reputation.

In order to survive in this highly competitive world, we simply must produce better-quality products at prices which are competitive. This not only means that we must improve the quality of our products, but use any or all of the new manufacturing processes and high-efficiency superabrasive tools to reach this goal. Producing quality products, competitively priced, delivered on time, and properly serviced throughout their lifetime, is the way to be competitive in the world marketplace. The by-products of quality products are increased customer satisfaction, reduced manufacturing costs, greater sales potential, more revenue, larger market share, and a more productive work environment (Fig. 6-3).

FUNCTIONAL COSTING

There is a definite need for factories to use the latest manufacturing processes and the best cutting tools to narrow the productivity gap that exists between U.S. and offshore workers. *Functional costing,* a new technique developed by Ingersoll Engineers and some of its clients, is a method undertaken by a team of engineering, manufacturing, and financial personnel to justify expenditures on new tools and equipment. This system differs from traditional factory costing systems in that it understands how and where money is spent in the factory. It has been used with great success to justify expenditures for major equipment purchases such as numerical control machines, flexible manufacturing systems, and superabrasive high-efficiency cutting tools.

Functional costing has three firm guidelines which allow a wide variety of formats and make the results easier to understand.

1. Group costs into four categories:
   a. Operator labor
   b. Factory overhead
   c. Production material

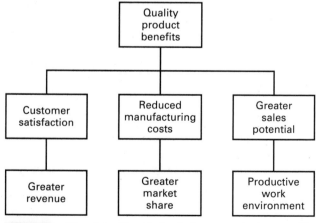

Fig. 6-3 The benefits derived from producing quality products.

d. Nonfactory costs

When done over an entire product line, this will give bottom-line figures that can be tied back to the official profit and loss (P&L) statement.

2. Group costs together if they are a result of a single decision:
   a. An item such as supervision may include both fixed and variable costs.
3. Assign meaningful names to new items:
   a. Do not use "incidental costs," etc.
   b. Use "cleaners and oils," for example.

If the engineering-manufacturing-accounting team is careful and thorough during the functional costing exercise, it can generate very accurate results of any factory upgrade such as capital equipment or high-efficiency cutting tools.

Tool Cost Savings

The addition of new superabrasive tools can have a very meaningful effect on reducing the manufacturing cost per part. Let us examine the spreadsheet shown in Fig. 6-4 to see what categories would be affected by the purchase of new high-efficiency superabrasive tools.

1. *Operator Labor*
   a. *Machine Operator.* There would be less operator cost per part because higher production rates result due to increased speeds and feeds.
   b. *Setup.* Less tool setup time is required because superabrasive tools stay sharp much longer than conventional tools.
   c. *Rework.* Since superabrasive tools produce better part accuracy and geometry, there is less need to rework parts.
   d. *Material Handling.* Since superabrasive tools produce accurate parts, the time required to handle parts is reduced.
2. *Factory Overhead*
   a. *Tooling.* The total cutting-tool cost per part manufactured for superabrasives is considerably less than for conventional tools because of their longer tool life and greater productivity.
   b. *Maintenance.* Less maintenance is required for superabrasive tools because of their superior

Generic Plant: Product Cost

COST AREA	MACHINED ASSEMBLIES			COST AREA	MACHINED ASSEMBLIES		
	ANNUAL COST (x 1000)	PRODUCT COST (PERCENT)	COST PER UNIT		ANNUAL COST (x 1000)	PRODUCT COST (PERCENT)	COST PER UNIT
OPERATOR LABOR:				NONFACTORY COSTS:			
MACHINE OPERATOR AND ASSEMBLY	$5250	20.9	$1.17	INVESTMENT CHARGES	$695	2.8	$0.15
SETUP	530	2.1	0.12	ADMINISTRATIVE DIVISION	560	2.2	0.12
REWORK	350	1.4	0.08	PERSONNEL DIVISION	420	1.7	0.09
OVERTIME PREMIUM	180	0.7	0.04	MANUFACTURING DEVELOPMENT	260	1.0	0.06
TRAINING	115	0.5	0.03	DATA PROCESSING DIVISION	255	1.0	0.06
NIGHT PREMIUM	100	0.4	0.02	ACCOUNTING	240	1.0	0.05
MISCELLANEOUS	45	0.2	0.01	CORPORATE CHARGES	235	0.9	0.05
MATERIAL HANDLING	25	0.1	0.01	MARKETING	175	0.7	0.04
SUBTOTAL	$6595	26.3	$1.48	PURCHASING DIVISION	170	0.7	0.04
FACTORY OVERHEAD:				QUALITY ASSURANCE	80	0.3	0.02
TOOLING	$2150	8.5	$0.48	RESEARCH AND DEVELOPMENT	20	0.1	0.00
MAINTENANCE	1275	5.1	0.28	LESS ALLOCATION OUT	-470	-1.9	-0.10
QUALITY ASSURANCE	585	2.3	0.13				
TEAM MANAGEMENT	540	2.1	0.12	NONFACTORY TOTAL	$2640	10.5	$0.58
PRODUCTION AND INVENTORY CONTROL	485	1.9	0.11				
PRESENT PRODUCT ENGINEERING	390	1.6	0.09				
DEPRECIATION	325	1.3	0.07	TOTAL COST	$25,161	100.0	$5.61
UTILITIES	300	1.2	0.07				
SUPPLIES	260	1.0	0.06				
SCRAP MATERIAL	175	0.7	0.04				
TAXES AND INSURANCE	135	0.5	0.03				
SALE OF SCRAP	80	0.3	0.02				
REWORK MATERIAL	24	0.1	0.01				
OTHER	22	0.1	0.00				
SUBTOTAL	$6746	26.7	$1.51				
PRODUCTION MATERIAL	$9180	36.5	$2.04	PROFIT	$3434	12.0	$0.76
FACTORY TOTAL	$22,521	89.5	$5.03	TOTAL SALES	$28,595		$6.37

Fig. 6-4 Functional costing considers all the costs involved in producing a product. *(Courtesy of Manufacturing Engineering)*

Jill Stevenson

hardness and wear resistance. On one production application, cemented carbide tools required maintenance *three times per day*, while superabrasive cutting tools needed maintenance *only once a month*.

c. *Quality Assurance.* Less time is required for inspection purposes because superabrasive tools consistently produce better part accuracy and geometry.

d. *Scrap Material.* There is less scrap material to be handled and stored. Therefore, there is a saving in material costs and also the cost of handling and storing of the scrap.

e. *Sale of Scrap.* Scrap material must be disposed of in some way and it is generally sold at very low prices. The administrative cost of selling scrap is greatly reduced if there is less scrap.

f. *Rework Material.* The amount of scrap is reduced when machining with superabrasive cutting tools; therefore, the amount of material which must be purchased to replace or repair parts is reduced.

3. *Production Material.* Because there is less scrap and rework necessary when machining with superabrasive cutting tools, there is a sizable reduction in material costs.

4. *Nonfactory Costs*

a. *Division Personnel.* Since better-quality parts are produced with superabrasive tools, there are fewer customer complaints to be handled. Field service costs are reduced because of better-quality parts.

b. *Quality Assurance.* There is less need for inspection and quality assurance because of the highly consistent performance of superabrasive tools.

Taking into account the various categories of the spreadsheet (Fig. 6-4) which could be affected by the use of high-efficiency superabrasive tools, even a small saving in each category would reduce the manufacturing cost per part considerably.

ANALYSIS OF GRINDING AND MACHINING COSTS

Many factors must be considered before arriving at the true cost of using any cutting tool. In the case of superabrasives, the productivity capabilities must be understood and taken into consideration in calculating grinding or machining costs. It has been usually assumed that the cost of the grinding wheel or cutting tool is the most important factor in the total cost equation. This section will show that while the cost of buying a conventional grinding wheel or cutting tool *may be low*, the cost of using these tools *may be very high*. The initial cost of a superabrasive wheel or tool blank may be *very high*; however, the cost of using these superabrasive tools may be *so low* that their initial cost is easily justified.

Analysis of Surface Grinding Costs

An example of surface grinding hardened high-speed steel tools has been selected in order to show the various factors involved in calculating total grinding costs. The bar chart in Table 6-1 shows the various cost factors to be considered for both aluminum oxide and CBN grinding wheels. Each of these cost factors are discussed as follows:

1. *Wheel Cost.* Table 6-1 shows that in both cases the cost of the grinding wheel is a small part of the total cost. The aluminum oxide wheel amounts to only 7 percent of the total cost, while the CBN wheel is around 10 percent.

2. *Grind Time to Remove Stock.* The time required to grind the part takes up the largest percentage of the time for both wheels. This must take into account the time cost while the wheel is in actual contact with the workpiece, but all the "deadtime" costs involved in the crossfeed, downfeed, and overpasses at each stroke reversal and at the workpiece sides while the wheel is not in contact with the workpiece must also be considered.

3. *Grind Time due to Wheel Breakdown.* When grinding *difficult-to-grind* (DTG) materials, grinding ratios are quite low because the grinding wheel is breaking down all the time that it is in contact with the workpiece. This breakdown during one pass over the workpiece means that less material is removed from the work than is fed down by the wheelhead. In the case of the aluminum oxide wheel, additional downfeeds and passes over the work were necessary to remove the required material, which adds greatly to the total grinding costs.

4. *Dressing Time.* Because of the rapid breakdown of the aluminum oxide wheel, it is necessary to often stop the grinding operation and redress the wheel to restore its sharpness and form.

5. *Gaging Time.* Because the aluminum oxide wheel breaks down so quickly, more time will have to be spent on measuring the workpiece to determine its actual size.

6. *Spark-out Time.* Because of the characteristics of the aluminum oxide wheel, it is necessary to spend more time to obtain the finish dimension and the required surface finish on a part.

7. *Wheel-change Time.* When grinding DTG workpieces, the aluminum oxide wheel wears rapidly and will be reduced to the minimum allowable diameter in a short period of time; therefore, there will be more downtime (less productivity) to change grinding wheels.

Considering all these factors, it can be seen that:

- The cost of the grinding wheel is a very small part of the overall grinding costs.

Table 6-1 **ECONOMIC EVALUATION OF CBN WHEELS**

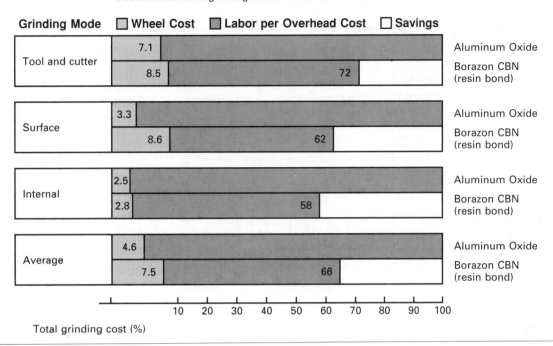

Documented Savings Range from 10 to 70% with BORAZON CBN

Grinding Mode □ Wheel Cost ■ Labor per Overhead Cost □ Savings

			Aluminum Oxide
Tool and cutter	7.1		Aluminum Oxide
	8.5	72	Borazon CBN (resin bond)
Surface	3.3		Aluminum Oxide
	8.6	62	Borazon CBN (resin bond)
Internal	2.5		Aluminum Oxide
	2.8	58	Borazon CBN (resin bond)
Average	4.6		Aluminum Oxide
	7.5	66	Borazon CBN (resin bond)

Total grinding cost (%)
10 20 30 40 50 60 70 80 90 100

(Courtesy of GE Superabrasives.)

- The use of a CBN wheel greatly reduces or eliminates costs due to wheel breakdown, which includes dressing time, gaging time, sparkout time, and wheel-change time.

- The true total cost due to rapid wheel wear makes up over half the cost of a grinding operation.

To arrive at a true picture of whether a superabrasive cutting tool will be economical, two factors must be considered for the total machining cost equation: the cost of using the cutting tool and the price of the cutting tool. Let us use a CBN grinding wheel to examine what factors must be considered to arrive at the total grinding costs per part produced.

Cost of Using the Wheel

1. The ability of a CBN grinding wheel to remove stock will determine the production rate and also the amount of labor and investment required to produce parts.
2. The ability of any wheel to remove stock is governed by the number of times that a wheel must be dressed in order to produce accurate work and a good surface finish.
3. The rate that a wheel wears will influence how often a worn wheel must be removed from a machine and a new wheel replaced and conditioned ready for grinding.

4. The wheel must be trued and stored in inventory before or after use, and this affects the total grinding cost.

Price of the CBN Wheel

When the cost of producing parts with aluminum oxide wheels is compared to CBN wheels, it will be found that this cost amounts to only about 5 to 10 percent of the total grinding cost per piece. Although CBN wheels are more expensive to buy, the cost of using these wheels is much lower than with aluminum oxide wheels because of their increased productivity.

The many factors that affect the true grinding cost per part are illustrated in Fig. 6-5. Although some factors deal with the type of abrasive, abrasive grit size, wheel bond, and wheel size, all have affected the grinding ratio or G ratio (the ratio of the amount of metal removed in proportion to the amount of wheel consumed) or metal-removal rate (MRR). Therefore, the selection of the proper wheel to suit the workpiece material and the grinding operation is very important to the efficient removal of metal.

The cost-benefit ratio of CBN wheels does not take into account the savings which occur in the reduction of scrap and the rework. The lack of surface damage to the workpiece as a result of consistently higher wheel sharpness, improved control over part accuracy and geometry, and reproducible surface finishes are also identifiable benefits of CBN grinding wheels.

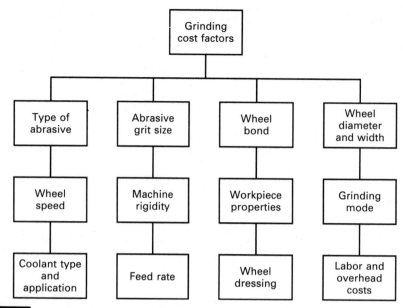

The factors which contribute to the total grinding cost per part produced.

Table 6-1 presents an economic evaluation of Borazon CBN and aluminum oxide grinding wheels under various grinding conditions. While the wheel cost per piece is generally a little higher for a CBN wheel than for an aluminum oxide wheel, labor and overhead costs are much lower because of the higher productivity of the CBN wheel. Therefore, the use of CBN wheels result in major savings in *total grinding cost per part*.

GRINDING AND MACHINING CASE HISTORIES

Superabrasives made of cubic boron nitride and diamond have found wide use in many grinding and machining applications. A few examples of their use, along with a cost comparison of how they perform in relation to conventional cutting tools, are presented in the following subsections.

Tool and Cutter Grinding

CBN wheels are superior to other abrasives for the dry grinding of milling cutters. Other abrasives wear rapidly when dry grinding, and the machine settings must be adjusted frequently to compensate for wheel wear and to maintain part geometry. When dry-grinding milling cutters with Borazon Type II, there is little or no wheel wear, which allows the machine operator to maintain tool geometry without dressing the wheel or adjusting the machine setting. Because of the CBN wheel's sharp cutting action and cooler grinding, little thermal damage occurs to the cutting edges of the tool, which improves the life of the milling cutter.

GRINDING COST COMPARISON. The following example of a field test between Borazon CBN Type II and an aluminum oxide wheel shows an example of the savings which are typical for a tool and cutter grinding operation. The operation (Table 6-2) involved the sharpening of $1\frac{1}{2}$- to 2-in. (38- to 50-mm)-diameter high-speed steel end mills.

RESULTS After grinding 1300 end mills, 42 percent of the original rim of the Borazon wheel remained. Therefore the estimated life of the wheel was 2260 end mills. The life of a comparable aluminum oxide wheel was normally averaging only 45 end mills. The Borazon wheel never required dressing during its use, while the aluminum oxide wheel had to be constantly dressed to maintain surface finish and tool geometry.

An average of 75 end mills were ground with aluminum oxide wheels during an 8-h shift. The Borazon wheel was able to grind 103 end mills during the same 8-h shift, a productivity increase of 40 percent. On the basis of these figures, it would require only 1 CBN wheel to grind 2260 end mills, while 50 aluminum oxide wheels would be required to grind the same number (Table 6-3A).

To arrive at the exact grinding cost for each end mill ground, the cost of the grinding wheel plus the labor and overhead rate must be considered. The grinding cost comparison for each piece between the CBN and the aluminum oxide wheel is shown in Table 6-3B.

Cubic boron nitride, although it costs more ini-

Table 6-2	GRINDING COSTS
	Test Conditions
Infeed	0.002–0.003 in. (0.05–0.07 mm) per pass
Wheel speed	3900 sf/min (1188 m/min)
Wheel type	$3\frac{3}{4}\times1\frac{1}{2}\times1\frac{1}{4}$ in. (95 × 38 × 32 mm) Type 11V9, $\frac{1}{16}$-in. rim, Borazon Type II 25VP, 60/80 mesh

Table 6-3A	CBN AND ALUMINUM OXIDE WHEEL GRINDING COMPARISON

Wheel Type	Wheel Life		Parts per 8-h Production	Parts per 40-h Production
	Parts	Hours		
Borazon Type II	2260	175.5	103	515
Aluminum oxide	45	4.8	75	375

Table 6-3B	CBN AND ALUMINUM OXIDE GRINDING COST COMPARISON

Wheel Type	Wheel Cost per Part	LOH* Cost per Part	Total Cost per Part
Borazon Type II	$0.15	$3.11	$3.26
Aluminum oxide	$0.11	$4.27	$4.38

*Labor and overhead rate based on $40.00/h.

tially, is a more productive wheel which lowers labor and overhead costs greatly; and the increase in improved productivity more than offsets the added cost—*CBN is cost-effective.*

Surface Grinding

Cubic boron nitride wheels are cost-effective for grinding a variety of metals which are generally considered difficult to cut. These include the following:

- Any carbon or alloy steel that has been hardened to Rc 50 (Rockwell "C" hardness of 50) or higher

- Extremely hard, abrasion-resistant cast irons and abrasion-resistant steel alloys

- Nickel-base and cobalt-base superalloys with a hardness of Rc 35 or higher

- Specialty metals such as Stellite

The benefits of surface grinding with CBN wheels are more consistent part accuracy, better surface quality, higher productivity, and lower grinding costs.

The cost-effectiveness of CBN grinding wheels is illustrated in Fig. 6-6, which is based on surface grinding operations on hardened steel broach bodies. Although wheel cost per tool is higher when grinding with a CBN wheel than with an aluminum oxide wheel, labor and overhead cost per tool is lower due to:

1. Reduced grinding time
2. Fewer wheel changes
3. Less wheel conditioning
4. Fewer stops for gaging and machine adjustments to compensate for wheel wear
5. Faster finishing cycles

The total grinding cost for each tool ground is reduced by 40 percent, making the CBN grinding wheel an excellent investment.

GRINDING COST COMPARISONS. The following field tests involving CBN and aluminum oxide wheels show examples of the cost savings possible for surface grinding operations.

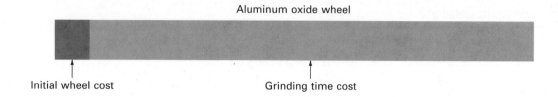

Aluminum oxide wheel

Initial wheel cost   Grinding time cost

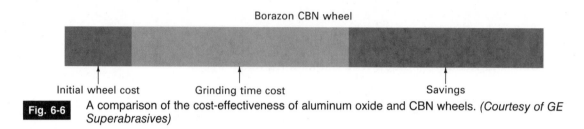

Borazon CBN wheel

Initial wheel cost   Grinding time cost   Savings

Fig. 6-6 A comparison of the cost-effectiveness of aluminum oxide and CBN wheels. *(Courtesy of GE Superabrasives)*

Example 1—Surface Grinding Magnetic Stud Assemblies

Material—Alnico V
Equipment—horizontal spindle grinder

See Table 6-4.

Table 6-4   **SURFACE GRINDING MAGNETIC STUD ASSEMBLIES**

Grinding Conditions

Per Table Load	Previous	Present
Abrasive	Aluminum oxide	Borazon CBN Type II
Wheel type	-A60F12V	Type 1A1, 80/100 mesh, 25 vol %
Wheel size	$7 \times \frac{3}{4} \times 1\frac{1}{4}$ in.	$7 \times \frac{3}{4} \times 1\frac{1}{4}$ in.
		$(17.8 \times 1.9 \times 3.2$ cm)
Wheel speed	6500 sf/min	6500 sf/min (33 m/s)
Table speed	30 ft/min	115 ft/min (35 m/min)
Crossfeed	0.015 in. per pass	0.020/pass (0.51 mm)
Downfeed	0.010 in.	0.010 in. (0.25 mm)
Coolant	Water-soluble oil, 1:50	Water-soluble oil, 1:50
Dressing	Each table load	None

Cost Evaluation

	Aluminum Oxide	Borazon CBN Type II
Time, min	14.8	7.7
LOH cost ($40.00/h)	$9.87	$5.13
Wheel cost	$0.20	$0.35
Total cost	$10.07	$5.48

Example 2—Surface Grinding Slotting Saws

Material—AISI—M2
Hardness—Rc 62
Equipment—vertical spindle grinder

See Table 6-5.

Table 6-5 **SURFACE GRINDING SLOTTING SAWS**

Grinding Conditions		
	Previous	**Present**
Abrasive	Aluminum oxide	Borazon CBN Type II
Wheel specifications	-A36FVOS, 6A2	80/100, 25 vol %, 6A2
Wheel size	8 × 3 in.	8 × 3, $\frac{1}{4}$×$\frac{1}{8}$ rim
Wheel speed	3770 sf/min	3770 sf/min
Downfeed	0.0005 in. per double pass	0.0005 in. per double pass
Traverse	12.5 ft/min	12.5 ft/min
Stroke length	22 in.	22 in.
Coolant	Water-soluble synthetic	Water-soluble synthetic
Dressing	Every two double passes	None

Cost Evaluation		
5 Parts/Load	**Aluminum Oxide**	**Borazon CBN Type II**
LOH Rate	**$40.00/h**	**$40.00/h**
Time, min	24	15
Labor cost	$16.00	$10.00
Wheel cost	$0.20	$0.35
Total cost	$16.20	$10.35

Internal Grinding

Cubic boron nitride wheels can reduce the cost of internal grinding steels and cast irons with a hardness of Rc 50 and above, and tough nickel-base and cobalt-base superalloys with a hardness of Rc 35 and above. Even though CBN wheels cost considerably more than conventional wheels, they outlast them by a wide margin when grinding hard, tough materials, making them cost-effective. They improve productivity tenfold or more by removing more stock in fewer passes.

Some of the benefits of using CBN wheels for internal grinding are:

- Less wheel consumption per part ground
- Less downtime (stops) for wheel changes and wheel conditioning
- Less downtime to measure part size and make machine adjustments
- Little or no metallurgical damage to the work surface

Example 1—Internal Grinding Blind Holes in Shoe Retainers

Material—Stentor (AISI 02)
Hardness—Rc 60
Equipment—internal grinder
Stock removal—0.010 to 0.012 in. (0.25 to 0.30 mm) in diameter

See Table 6-6.

Table 6-6 **INTERNAL GRINDING BLIND HOLES**

Grinding Conditions		
	Previous	**Present**
Abrasive	Aluminum oxide (white)	Borazon CBN Type II, 80/100
Wheel type	-A60KV5, Type I	Type 1A1, 25 vol %
Wheel size	$1\frac{1}{2}$×$1\frac{1}{2}$×$\frac{5}{8}$ in.	$1\frac{1}{2}$×$\frac{1}{4}$×$\frac{5}{8}$–$\frac{1}{8}$ in. rim
Wheel speed	4400 sf/min	4400 sf/min (22.3 m/s)
Table speed	4 in./min	8 in./min (20.3 cm/min)

Table 6-6 (*continued*)   INTERNAL GRINDING BLIND HOLES

Grinding Conditions

	Previous	Present
Stroke length	$\frac{1}{8}$ in.	$\frac{1}{8}$ in. (3.18 mm)
Infeed (each direction)	0.0002 in.	0.00015 in. (0.004 mm)
Coolant	Immunal—517 (synthetic)	Immunal—517 (synthetic)
Dressing	Each part	None
Finish	10 RMS	7 RMS

Cost Evaluation

	Aluminum Oxide	Borazon CBN Type II
LOH Rate	$40.00/h	$40.00/h
Minutes required/part	4.3	3.3
Labor cost/part	$2.87	$2.20
Wheel cost/part	$0.015	$0.003
Total cost/part	$2.885	$2.203

Example 2—Internal Grinding Wear Plates

Material—52100
Hardness—Rc 58 to 62
Stock removal—0.010 to 0.012 in. (0.25 to 0.30 mm) in diameter
Equipment—internal grinder

See Table 6-7.

Table 6-7   INTERNAL GRINDING WEAR PLATES

Grinding Conditions

	Previous	Present
Abrasive	Aluminum oxide (white)	Borazon CBN Type II, 80/100
Wheel type	-A60K5V, Type I	Type 1A1, 25 vol %
Wheel size	$1\frac{1}{2} \times 1\frac{1}{2} \times \frac{5}{8}$ in.	$1\frac{1}{2} \times \frac{1}{4} \times \frac{5}{8} - \frac{1}{8}$ in rim
Wheel speed	4400 sf/min	4400 sf/min (22.3 m/s)
Table speed	10 in./min	30 in./min (76.2 cm/min)
Stroke length	$\frac{1}{8}$ in.	$\frac{1}{4}$ in. (6.35 mm)
Infeed (each direction)	0.0002 in.	0.0003 in. (0.008 mm)
Coolant	Immunal—517 (synthetic)	Immunal—517 (synthetic)
Dressing	Every 3 parts	None
Stock-removal rate	0.67 in.^{3}/h	1.6 in.^{3}/h

Cost Evaluation

	Aluminum Oxide	Borazon CBN Type II
LOH rate	$40.00/h	$40.00/h
Minutes required/part	2.5	1.7
Labor cost/part	$1.67	$1.13
Wheel cost/part	$0.002	$0.002
Total cost/part	$1.672	$1.132

Jig Grinding

Continuous-path numerical control (NC) jig grinding requires grinding abrasives that last a long time, retain their shape, produce good surface finishes, and maintain size and form without thermal damage to the workpiece. One factor which is impossible to program is wheel wear. If the wheel loses shape, size, or stock-removal capability while making a pass, an inaccurate form will be produced.

Cubic boron nitride wheels make practical the precise high stock removal that is required for economic NC jig grinding. Some of the main advantages that CBN wheels have over conventional wheels for jig grinding are:

- Longer wheel life

- Less wheel maintenance

- More positive size control

- More consistent surface finishes

- Less spindle deflection

- Cooler grinding

- Faster stock removal

- Reduced jig grinding costs

ECONOMICS OF JIG GRINDING WITH SUPERABRASIVES. Superabrasive wheels outperform conventional abrasive wheels by a wide margin when jig grinding hardened steels and DTG materials. They are widely used for jig grinding because they stay sharp and free-cutting; therefore, there is less spindle deflection, which results in precise stock removal. The longer life of CBN wheels, shorter grinding cycles, and improved part quality result in lower jig grinding costs. Superabrasive diamond wheels are used almost exclusively instead of silicon carbide wheels for the jig grinding of cemented carbide components and tooling.

Since superabrasive jig grinding wheels give long life under severe production conditions, they are much more cost-effective than the conventional wheels used previously.

Honing

Superabrasive stones are now honing many materials that until recently have resisted cutting by any method. Most of the popular tool steels, such as the A, D, H, M, and T series, can now be honed as easily as ordinary hardened steels. Even difficult materials such as Stellite, Inconel, Ferrotic, Alnico, and Alumina can be honed routinely, quickly, and with predictable results.

Some of the advantages of superabrasive honing stones are:

- Increased productivity

- Lower abrasive cost

- Improved part geometry

- Less machine and tooling wear

- Less coolant contamination

- Noise pollution (squeal) reduced

ECONOMIC EVALUATION. Honing costs can be divided into two categories:

1. *Abrasive cost.* The cost of a set of stones divided by the parts that can be honed with the stones
2. *Labor and Overhead Cost.* Labor and overhead cost per hour divided by the number of parts honed per hour

Typically the cost breakdown for a mild steel honing operation with aluminum oxide would be 20 to 25 percent stone cost and 75 to 80 percent labor and overhead cost.

	ALUMINUM OXIDE	
	Abrasive Cost	Labor and Overhead Cost
Total Cost	25%	75%

A typical CBN cost breakdown would be 30 percent stone cost, 55 percent labor and overhead cost, and 15 percent savings.

	CUBIC BORON NITRIDE		
	Abrasive Cost	Labor and Overhead Cost	Savings
Total cost	30%	55%	15%

In the above example there is a slight increase in abrasive cost with CBN but a substantial reduction in honing time. In other cases, a CBN cost breakdown can be shown as:

	CUBIC BORON NITRIDE		
	Abrasive Cost	Labor and Overhead Cost	Savings
Total cost	15%	50%	35%

Both abrasive cost and labor and overhead cost are lower for CBN than those of conventional aluminum oxide honing.

Example 1—Honing Drill Bushings

Material—M5
Hardness—Rc 62
Stock removal—0.004 in. (0.10 mm)
Equipment—honing machine

See Table 6-8.

Table 6-8	HONING DRILL BUSHINGS	
	Cost Evaluation	
	Silicon Carbide	**Borazon CBN**
LOH rate	$40.00/h	$40.00/h
Stone life	62 parts	958 parts
Minutes required/part	8 min	1 min
Labor cost/part	$5.33	$0.67
Stone cost/part	$0.01	$0.03
Total cost/part	$5.34	$0.70

(Courtesy of GE Superabrasives.)

Machining with Polycrystalline Tool Blanks

Diamond and CBN superabrasive cutting-tool blanks are used by industry throughout the world to improve the productivity of machining operations on difficult-to-cut (DTC) materials. They consist of a layer of diamond or CBN bonded to a cemented carbide substrate. Polycrystalline diamond tools (PCD) are used for turning and milling operations on nonferrous and nonmetallic materials, primarily where the workpiece is abrasive. Polycrystalline cubic boron nitride (PCBN) tools are used for machining (turning and milling) hard, DTC, and hard abrasive ferrous metals.

Some of the chief advantages of diamond and CBN blank tools are:

- They stay sharp and cut efficiently through long production runs.

- Higher cutting speeds and material-removal rates are possible with less tool wear.

- They require few tool changes and less production stoppage.

- Consistently smoother surface finishes are produced.

- There is better control over workpiece shape and size.

ANALYSIS OF MACHINING COSTS. Examples of various turning and milling operations with PCBN and PCD tool blanks are shown in Table 6-9. The bar chart in this table compares the cost of machining various materials with tungsten carbide and superabrasive tool blanks.

1. *Tool Cost.* Table 6-9 shows that in both cases the cost of the cutting tool is a small part of the total cost of producing a part. The tungsten carbide tool

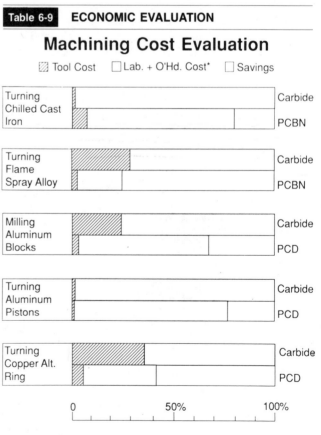

Table 6-9	ECONOMIC EVALUATION

Machining Cost Evaluation

*Assumed @ $30/hr.

averages only about 7 percent of the total cost, while the superabrasive tool is around 10 percent.

2. *Machining Time to Remove Stock.* The time required to machine a part takes the largest percentage of the time for both cutting tools. This must include the cost of all the dead time when the tool is not in contact with the workpiece, such as setting cuts, gaging, and tool conditioning.

3. *Machining Time due to Tool Breakdown.* When machining DTC materials, the metal-removal ratios are lower because the cutting tool is wearing all the time it is in contact with the workpiece. This breakdown, during each pass over the workpiece, means that less material is removed than is set for on the machine. Conventional cutting tools require additional passes to remove the required material, while the wear on superabrasive tools is so small that additional passes are not necessary.

4. *Gaging Time.* Because conventional cutting tools break down quickly, more time will have to be spent on measuring the workpiece to determine its actual size. With superabrasive tools, gaging time is greatly reduced and almost eliminated on most jobs.

5. *Tool Change or Reconditioning Time.* When machining DTC material, conventional cutting tools wear rapidly and must be replaced or reconditioned much more often than superabrasive tools. On one machining example, the tungsten carbide tool required replacing three times a day while the superabrasive tool required replacement only once a month.

Considering all these factors, it can be seen that:

• The cost of the cutting tool is a very small part of the overall machining cost.

• The use of superabrasive cutting tools greatly reduces or eliminates costs due to tool wear, gaging time, and resetting the machine position.

• The true total cost due to rapid tool wear makes up over half the cost of a machining operation.

ECONOMIC EVALUATION. Both polycrystalline diamond and cubic boron nitride blank tools can be used for turning and milling operations but on different types of material. Industry has found these superabrasive cutting tools among the most effective tools for production cost reduction and product improvement (Table 6-9). These tools outperform carbide cutting tools by a wide margin and are the most cost-effective tools available today. Although superabrasive cutting tools cost more to buy, they cost less to use as shown in the following examples.

1. *PCD tool examples:*
   a. Turning operation—boring (see Table 6-10)
   b. Milling operation (see Table 6-11)
2. *PCBN tool examples:*
   a. Turning operation (see Table 6-12)
   b. Milling operation (see Table 6-13)

Table 6-10	BORING ALUMINUM ALLOY GEAR CASES	
	Machine operation	Boring
	Part description	Gear case
	Material specification	Aluminum alloy, Type 380

Machining Conditions	Tungsten carbide	Compax* Blank Tool
Depth of cut	0.030 in. (0.76 mm)	0.030 in. (0.76 mm)
Cutting speed	2800 sf/min (853 m/min)	2800 sf/min (853 m/min)
Feed	6 in./min (152 mm/min)	6 in./min (152 mm/min)
Coolant	Soluble oil	Soluble oil
Tool Geometry		
Nose radius	0.031 in. (0.78 mm)	0.031 in. (0.79 mm)
Side relief	20°	20°
Back rake	0	0
Economics		
Tool cost	X	23X
Cutting edges per tool	3	1
Pieces per edge	300	12,000–14,000
Pieces per tool	900	12,000–14,000

*Trademark of General Electric Company, USA.

Table 6-11   FACE MILLING TYPE 380 ALUMINUM ENGINE BLOCKS

Operation	Face milling engine block
Work material	Type 380 aluminum
Cutter	10 in. (250 mm) face milling, 5 inserts
Axial rake	5°
Radial rake	5°
Nose radius	0.031 in. (0.79 mm)
Speed	10,000 sf/min (3048 m/min)
Feed rate	100 in./min (30.48 m/min)
Depth of cut	0.100 in. (2.54 mm)
Coolant	Dry

Tool Material	Tungsten Carbide	Compax Blank Tool
Finish, in.	30–35	20
Insert cost, $	X	26.3X
Edges/insert	4	1
Edge life, pieces	40	18,000
Insert life, pieces	160	18,000

Table 6-12   TURNING CHILLED CAST-IRON ROLLS (HRc 62) WITH PCBN TOOLS

Machining operation	Turning
Part description	Chilled cast-iron roll
Material specification	HRc 61–62
Type of tool replaced	Tungsten carbide and grinding operation

Machining Conditions	Replaced Tool	BZN Compact Tool
Depth of cut	0.125 in. (3.17 mm)	0.100 in. (2.54 mm)
Cutting speed	50 sf/min (15.2 m/min)	300 sf/min (91.4 m/min)
Feed/revolution	0.020 in./rev (0.508 mm)	0.012–0.20 in./rev (0.30–0.50 mm)
Coolant	Compressed air	Compressed air
Tool Geometry		
Style of insert	Rectangular insert $1\frac{1}{4}\times\frac{3}{4}\times\frac{1}{2}$ in. (31.2 × 18.7 × 12.5 mm)	BRNG-43

Economics
Productivity of the turning operation more than doubled, and the finish grinding operation has been eliminated resulting in a major cost savings.

Size was held to within 0.005 in. (0.12 mm) from one end of the roll body to the other.

Table 6-13   CYLINDER HEAD FACE

Machining operation—milling head face
Part description—gray cast iron
Material specification—190–250 BHN
Cutter—10-in. (250-mm) diameter; 32 inserts double
  negative face milling
Inserts—SNG-632 (15° × 0.005 in. land)
Speed—3100 sf/min (950 m/min)
Feed—0.005 in. (0.12 mm)/rev
Depth of cut—0.020 in. (0.5 mm)
Coolant—dry
Results—17,000 pieces/corner (CBN)
    —1900 pieces/corner (SiN)

Roadblocks to Justification

1. Compare the labor costs of the 1920s with those of the 1980s.
2. List three reasons why superabrasive tools can be justified on quality products alone.

New Tool Advantages

3. List eight advantages of high-efficiency superabrasive cutting tools.

Functional Costing

4. Name the three cost areas of functional costing.

Economic Performance

5. What two factors must be considered for the total machining cost equation regarding superabrasive tools?
6. What four factors affect the cost of using a CBN grinding wheel?

Grinding and Machining Case Histories

7. Why are CBN wheels superior to other wheels for the dry grinding of milling cutters?
8. Name five reasons why the labor and overhead cost per tool is lower when surface grinding with CBN wheels.
9. For what type of work materials are the following tools used:
   a. polycrystalline diamond?
   b. polycrystalline cubic boron nitride?
10. List five advantages of diamond and CBN blank tools.

CHAPTER SEVEN

CBN Grinding Applications

The trend in industrial production toward higher productivity with the same or increased part accuracy, using harder-to-machine materials naturally leads to the increasing use of grinding to solve many machining problems. Conventional abrasives, although suitable for many applications, cannot meet the production requirements, and in many cases, grinding with superabrasives such as diamond or CBN wheels seems to be the solution. The application of superabrasive technology is not limited to hardened materials, but is expanding into more and more applications in the machining of annealed or mild steels.

Previously, there were no grinding processes which could be substituted for traditional machining processes of turning, milling, and external broaching. The development of machine tools specifically designed for superabrasive grinding wheels and of the proper wheels for each application has made possible the economic machining of hard materials at high material-removal rates.

Cubic boron nitride (CBN) wheels are used for the grinding of hard, difficult-to-machine ferrous metals. They remove material at lower temperatures than those which occur with conventional abrasives under identical process conditions. Because of the lower grinding heat generated by CBN grinding wheels, there is little or no thermal damage occurring to the surface of the workpiece. Therefore, cutting tools and the parts ground with CBN wheels generally have a longer useful life. Material-removal rates are so high that they can replace lathe, milling machine, and conventional grinding operations.

Initially, CBN wheels were used on tool and cutter grinders in toolrooms for sharpening cutting tools such as end mills, milling cutters, broaches, etc. As their success continued to grow, they were used in other types of grinders. Today they are widely used on all types of surface grinders, cylindrical and internal grinders, centerless grinders, and special-purpose grinders.

Surface Grinding

Surface grinding generally refers to the production of flat, contoured, and irregular surfaces on a piece of work which is fed into a rotating grinding wheel in a single plane. The required form can be produced on a part through a combination of a successive number of grinding-wheel downfeed and crossfeed increments, or in one pass by creep-feed grinding.

Cubic boron nitride grinding wheels are designed to grind hardened tool and die steels more efficiently than either diamond wheels or aluminum oxide wheels. When CBN wheels are used correctly, the result is increased productivity, improved work quality, and reduced grinding costs. CBN wheel wear is slow and uniform, and wheel life is long. This is especially apparent when grinding difficult-to-grind (DTG) materials at high material-removal rates. The low wheel wear makes it easier to control part size, and there is no need to stop the machine frequently to check part size and adjust the grinder to compensate for wheel wear. Also, it is not necessary to dress the wheel frequently to maintain straightness or form.

When CBN grinding wheels are used properly, the grinding operation is relatively cool and there is almost no danger of burning the workpiece. Because of the absence of thermal damage, cutting tools ground with CBN wheels have a much longer tool life. The cool-grinding characteristics of CBN wheels make it easier to control straightness when grinding thin parts. When grinding heat is excessive, as is often the case with aluminum oxide wheels, residual stresses may be set up that cause parts to warp when released from the magnetic chuck after grinding (Fig. 7-1).

OBJECTIVES

After completing this section you should be able to:

1. Select the proper wheel for the grinding operation and the type of material being ground
2. Mount, true, and dress the CBN wheel
3. Grind in various modes on a surface grinder
4. Recognize common grinding problems and try various remedies to correct the same

TYPES OF SURFACE GRINDERS

There are four basic types of surface grinding machines, all of which have some way of holding the piece of work and passing it against a revolving grinding wheel.

The horizontal spindle grinder with a reciprocating table (Fig. 7-2) is the most common surface grinder used for toolroom work. The work, held on a magnetic chuck or suitable holding device, is reciprocated (moved back and forth) under a revolving grinding wheel which is fed down to provide the desired depth of cut. Crossfeed is obtained by the traverse movement of the table.

The horizontal spindle grinder with a rotary table (Fig. 7-3) is generally used for grinding flat circular parts. The surface pattern produced makes it especially suited for grinding parts which must rotate in contact with each other. The work is held on the magnetic chuck of a rotary table and passed under a revolving grinding wheel. Feed is obtained by the traverse movement of the wheelhead.

The vertical spindle grinder with a rotary table (Fig. 7-4) grinds with the face of the grinding wheel. The surface pattern produced appears as a series of intersecting arcs. Vertical spindle grinders have high material-removal rates and are generally considered as the most efficient and accurate form of grinder for the production of flat surfaces.

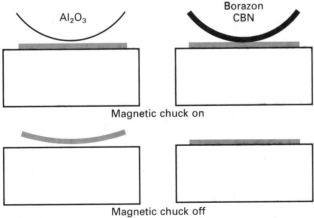

Fig. 7-1   CBN wheels are cool-cutting and do not set up residual stresses that may cause parts to warp when released from the magnetic chuck. *(Courtesy of GE Superabrasives)*

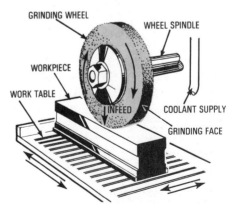

Fig. 7-2   The horizontal spindle surface grinder with a reciprocating table is the most common grinder used for toolroom work. *(Courtesy of the Carborundum Company)*

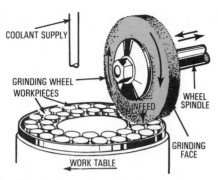

Fig. 7-3 The horizontal spindle (surface) grinder with a rotary table is particularly useful for grinding parts which must rotate in contact with each other. *(Courtesy of the Carborundum Company)*

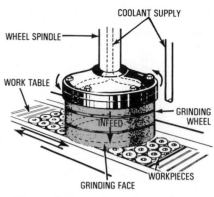

Fig. 7-5 The vertical spindle surface grinder with a reciprocating table grinds on the face of the wheel and is capable of taking heavy cuts. *(Courtesy of the Carborundum Company)*

The vertical spindle grinder with a reciprocating table (Fig. 7-5) grinds on the face of the wheel while the work is moved back and forth under the revolving wheel. Because of its vertical spindle and larger area of contact between the wheel and work, this grinder is capable of taking very heavy cuts (as much as ½ in. or 13 mm) in one pass.

TYPES OF MATERIAL GROUND

Cubic boron nitride wheels can grind all hardened tool and die steels and superalloys efficiently. CBN wheels resist breakdown better than do aluminum oxide wheels when grinding these hard steels and therefore improve productivity and reduce the grinding cost per piece.

The following types of steels are most effectively surface-ground:

1. *Hardened Tool and Die Steels.* CBN wheels resist wheel breakdown when grinding tool and die

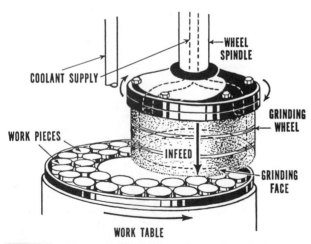

Fig. 7-4 The vertical spindle surface grinder with a rotary table grinds with the face of the wheel and is probably the most efficient form of grinder. *(Courtesy of AVCO Bay State Abrasives)*

steels hardened to Rc 58 to 60 and higher. Among the steels successfully ground with CBN wheels are M-2, M-4, M-15, T-1, T-9, T-15, A-7, and D-7.

2. *Hardened Steels and Cast Irons.* Almost any carbon or alloy steel hardened to Rc 50 or higher, and extremely hard, abrasion-resistant cast steel alloys can be successfully ground with CBN wheels. This includes stainless and bearing steels.

3. *Wear-Resistant Alloys and Superalloys.* CBN wheels are highly recommended for surface grinding hardened Stellites and wear-resistant alloys. They are especially well suited to the grinding of tough, gummy nickel-base and cobalt-base superalloys with hardness of Rc 35 and above.

The grindability or ease at which metals can be ground is greatly influenced by the composition of each metal. The addition of elements such as tungsten, molybdenum, chromium, nickel, cobalt, etc., impart certain characteristics to steel; however, they also make them more difficult to grind. Figure 7-6 lists the grindability of various metals which can be effectively ground with CBN wheels.

PREPARING THE GRINDER

Grinding with CBN wheels is very different from grinding with aluminum oxide wheels, and certain precautions must be taken so that the advantages that CBN wheels offer will be realized. Chapter 5 extensively covers the grinder characteristics, wheel selection, mounting, and truing and dressing procedures which are critical to the use of CBN wheels. Be sure to see Chapter 5 before attempting to use a CBN wheel.

The following summarizes some key points in the use of CBN wheels for surface grinding:

1. The grinder must have good rigidity in order to handle the high grinding forces.
2. Be sure that the grinder has enough spindle horsepower to maintain steady spindle speeds.

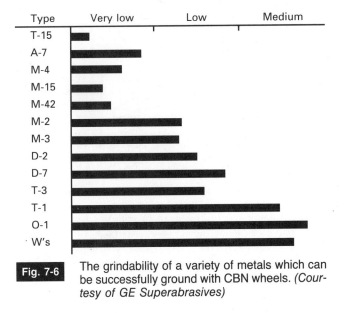

Type	Very low	Low	Medium
T-15			
A-7			
M-4			
M-15			
M-42			
M-2			
M-3			
D-2			
D-7			
T-3			
T-1			
O-1			
W's			

Fig. 7-6 The grindability of a variety of metals which can be successfully ground with CBN wheels. *(Courtesy of GE Superabrasives)*

3. Select the proper wheel to suit the workpiece material and type of grinding operation (Table 7-1).
   a. For most jobs, use resin-bond wheels.
   b. For grinding complex, deep forms, use electroplated wheels. Chapter 4 gives a detailed explanation of selecting CBN wheels.
4. Mount the wheel securely on the grinder spindle so that its runout is only 0.001 in. (0.02 mm) or less.
5. True the CBN wheel using the proper truing device and procedure recommended for each wheel type. (Electroplated wheels do not require truing.)
6. Always dress a wheel after truing to expose abrasive crystals and make the wheel grind efficiently. (Electroplated wheels do not require dressing.)

7. Use the proper speed and feed rates when surface grinding with CBN wheels.
8. Always use a good supply of cutting fluid properly directed to the wheel-work interface when surface grinding.

SURFACE GRINDING WITH CBN WHEELS

For most standard surface-grinding operations, resin-bond CBN abrasive wheels with 100 concentration are recommended. A 100-concentration wheel contains approximately 72 carats of abrasive per cubic inch (4.4 carats per cubic centimeter) of wheel rim material. Once the right CBN wheel has been mounted on the machine and correctly trued and dressed, it is ready for efficient grinding. To obtain the full benefits that a CBN wheel can provide, it is necessary to establish the proper operating conditions (Table 7-2).

1. *For Best Results, Grind Wet.* Because CBN wheels are used to grind DTG materials at high material-removal rates, a continuous flow of coolant is required to reduce grinding friction and carry away heat. Dry grinding is not recommended because the grinding heat may damage the workpiece. If it is necessary to grind dry, reduce the spindle speed, depth of cut, and table speed, to reduce the possibility of burning the workpiece.
2. *Use Soluble-Oil Coolants.* CBN wheels remove work materials by producing chips. Grinding pressures are heavy, and coolants should have good lubricity to prevent chips from sticking to the abrasive crystals. CBN wheel life and stock-removal capabilities are greatly improved by using a soluble oil coolant (Fig. 7-7).

Table 7-1 BORAZON CBN ABRASIVE APPLICATIONS

Applications	Bond System	Recommended Product	Workpiece Material
Tool and cutter grinding	Resin	Type II, 560	Hardened steels, difficult to grind steels, wear-resistant alloys, and superalloys
Surface grinding			
Jig grinding	Electroplated	Type I, 500, 570	
Form grinding	Vitrified, metal	Type I, 510, 550	
Internal-diameter grinding	Resin	Type II, 560	
	Electroplated	Type I, 500, 570	
	Vitrified, metal	Type I, 510, 550	
Hob grinding	Resin, metal	Type II, 510, 550, 560	
Vertical spindle rotary table grinding	Metal	510, 550	
Honing	Metal	510, 550	Hardened steel, mild steel
Creep feed grinding	Vitrified, metal, resin, electroplated	Type I, Type II, 500, 510, 550, 560, 570	Mild to hardened steels and superalloys
Superabrasive machining	Vitrified, metal, resin, electroplated	550, 560, 570	Mild and hard steels, nickel- and cobalt-base alloys
Lapping and polishing		Micron powder	Hardened steels

Table 7-2 SURFACE-GRINDING GUIDELINES

Guidelines	Comments
1. Grind difficult-to-grind (DTG) materials only	—CBN wheels are most effective for grinding hardened tool and die steels, other hardened steels, cast irons, Stellite, and tough superalloys —They are less effective when grinding easy-to-grind (ETG) materials
2. Use a rigid machine in good condition	—Good machine rigidity helps to prevent vibration and chatter —This prolongs CBN wheel life and improves surface finishes
3. Make sure that the machine has adequate horsepower	—To avoid spindle slowdown, spindle horsepower should be at least as high as the minimum shown in Table 5-1
4. For most jobs, use resin-bond CBN wheels	—Resin-bond wheels offer a good combination of high productivity and long wheel life —Use electroplated wheels for grinding deep or complex forms
5. Use normal-sized wheels	—Generally, small-diameter or narrow wheels wear fast —It is more economical to specify normal-sized CBN wheels
6. For most jobs, use 100-concentration wheels	—In most cases, 100-concentration wheels are a cost-effective choice, providing excellent wheel life
7. Use the coarsest abrasive grit that will produce the desired finish up to 100 mesh	—Usually, coarse abrasive grit sizes permit higher material-removal rates —This results in higher productivity than is possible with fine abrasive grit sizes
8. Mount the wheel securely	—Secure mounting will minimize chatter and vibration and promote grinding accuracy
9. True the wheel after mounting	—An out-of-round wheel will pound the workpiece, shorten wheel life, and produce poor surface finishes —True according to the directions that come with the wheel
10. Condition the wheel after truing	—Truing will glaze the wheel, and it will not cut —Condition according to the directions that come with the wheel
11. For best results, grind wet	—Use a 5 or 10 percent heavy-duty soluble oil —Apply the coolant stream directly to the wheel–workpiece interface
12. Wet-grind at 5000–6500 sf/min (25–33 m/s)	—Speeds in this range optimize wheel performance
13. Use high table speeds	—For maximum productivity, table speeds as high as the maximum grinder speed can be used
14. Use large crossfeed increments	—About ¼ to ½ wheel width feed increments are recommended
15. Use correct downfeeds	—For hardened steels such as M-2 and T-1, use 0.001 to 0.002 in. (0.02 to 0.05 mm) per pass —For DTG hardened steels such as M-4 and T-15, use 0.0005 to 0.001 in. (0.01 to 0.02 mm) per pass
16. Reduce feeds for finishing passes	—It is not necessary to condition CBN wheels before finishing passes

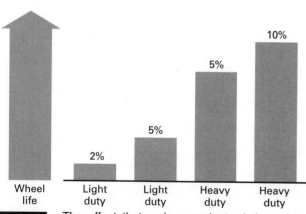

Fig. 7-7 The effect that various coolant-solution mixes have on CBN wheel life. *(Courtesy of GE Superabrasives)*

a. For general-purpose grinding, a 5 percent solution of light-duty soluble oil is suitable.

b. For superior results and longer wheel life, use a 5 to 10 percent solution of heavy-duty soluble oil.

c. For grinding difficult superalloys, sulfurized or sulfochlorinated mineral oils are recommended.

3. *Apply Coolant at the Work-Wheel Interface.* For best results, apply a good supply of coolant to prolong wheel life, reduce grinding heat, and improve surface finishes.

a. The coolant nozzle should be set parallel to the direction of table reciprocation, within ¼ in. (6 mm) of the work surface, and as close as possible to the wheel (Fig. 7-8).

b. Provide extra shielding around the coolant nozzle to confine the coolant spray.

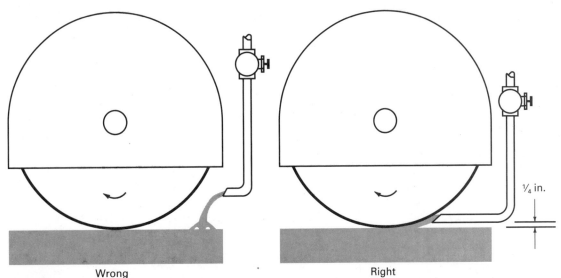

Wrong   Right

Fig. 7-8 For best results, the coolant nozzle should be set as close as possible to the top of the work surface and the wheel face. *(Courtesy of GE Superabrasives)*

c. Use a dummy block at the right end of the workpiece so that the surface being ground is always wet (Fig. 7-9).

4. *Use Proper Speeds and Feeds When Surface Grinding.*

a. *Wheel surface speeds* should be in the range of 5000 to 6500 sf/min (25 to 33 m/s). At speeds within this range, longest wheel life and finest surface finishes are obtained.

b. *High table speeds,* up to the maximum speed of the grinder, result in the best grinding performance. At high table speeds, material-removal rates are high, thus increasing the number of pieces ground and reducing labor and overhead costs per piece.

c. *Large crossfeed increments,* one-fourth to one-half of the wheel width, are recommended for general-purpose grinding (Fig. 7-10). Use smaller crossfeed increments for finishing cuts.

d. *Correct downfeeds* for CBN wheels range from:
   (1) 0.001 to 0.002 in. (0.02 to 0.05 mm) per pass for easy-to-grind (ETG) materials such as M-2 and T-1 steels.
   (2) 0.0005 to 0.001 in. (0.01 to 0.02 mm) per pass for DTG steels such as M-4 and T-15.

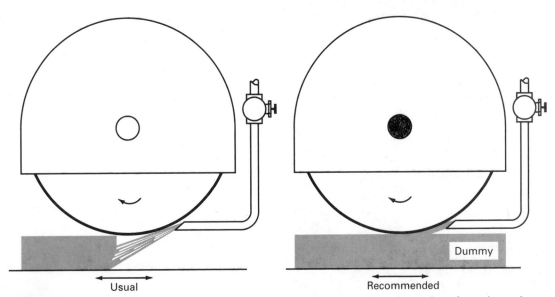

Usual   Recommended

Fig. 7-9 A dummy block at the right end of the workpiece ensures that the work surface always has a good supply of coolant. *(Courtesy of GE Superabrasives)*

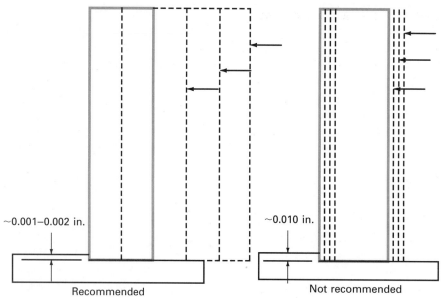

~0.001–0.002 in.   ~0.010 in.

Recommended   Not recommended

Fig. 7-10 Large crossfeeds help to extend CBN wheel life in surface grinding operations. *(Courtesy of GE Superabrasives)*

Grinding Flat Surfaces on a Horizontal Spindle Reciprocating Surface Grinder

The most common operation performed on a surface grinder is that of grinding flat surfaces. Figure 7-11 shows a large die made of Neatro (AISI M-4) Rc-62 steel which must be reground. This job will be used to show the procedures which should be followed to set up the machine, wheel, and grind flat surfaces. Be sure that the grinder bearings are in good condition before attempting to grind with cubic boron nitride (CBN) wheels (Table 7-3).

PROCEDURE: REGRINDING THE DIE

1. Select a 100-concentration CBN resin-bond wheel to grind the M-4 steel.

Fig. 7-11 Regrinding a die made of M-4 steel with a CBN wheel. *(Courtesy of GE Superabrasives)*

2. Mount the CBN wheel securely on the grinder spindle; *do not use wheel blotters.* (See Section 5 for complete instructions on wheel mounting procedures.)
3. True and dress the wheel correctly to ensure that the CBN wheel will perform properly. *Unless this operation is performed correctly, the CBN wheel will not cut.* (Refer to Chapter 5 for truing and dressing procedures.)

MOUNTING THE WORKPIECE

4. Remove any burrs from the magnetic chuck with an oilstone and thoroughly clean the chuck surface.
5. Remove any burrs and clean the bottom of the die.
6. Place a piece of smooth paper, no more than 0.002 in. (0.005 mm) thick and slightly larger than the die base, in the center of the magnetic chuck.
7. Mount the die on the paper and energize the magnetic chuck.

SETTING SPEEDS AND FEEDS

8. Adjust the table reverse dogs so that the center of the grinding wheel clears each end of the die surface to be ground by about 1 in. (25 mm).
9. Set the table crossfeed:
   a. Rough grinding—one-quarter to one-half the wheel width
   b. Finish grinding—smaller crossfeed increments
10. Set the table speed rate from 50 to 100 ft/min (15 to 30 m/min).

Table 7-3 CBN SURFACE GRINDING CHECKLIST

Problem	Possible Cause	Remedy
1. Poor finishes	Chatter, vibration	—Check to see that wheel mounting is tight —Check condition of spindle (replace bearings if necessary) —Wheels over 12 in. (305 mm) in diameter should be dynamically balanced
	Excessive crossfeed or downfeed Wrong coolant Poor placement of coolant Abrasive grit size too coarse Spindle rev/min too low	—Reduce feeds for finishing passes —Use 5–10 percent heavy-duty soluble oil —Apply directly to wheel–workpiece interface —Use wheel with smaller abrasive grit size —Grind at 5000–6500 sf/min (25–33 m/sec) wheel speed (*Note:* Inadequate horsepower may cause spindle to slow down.)
2. Burning, excessive heating	Wheel glazed or loaded Wrong coolant Poor placement of coolant Excessive material-removal rate	—Condition or recondition wheel —Use 5–10 percent heavy-duty soluble oil —Apply directly to wheel-workpiece interface —Reduce downfeed and/or crossfeed
3. Wheel won't grind	Wheel glazed or loaded	—Condition wheel (*Note:* Always condition wheel after truing.)
4. Short wheel life	Undersized wheel Exceeding recommended grinding parameters Wheel not true	—Always specify normal-sized wheel —Double-check wheel recommendations —True wheel

SETTING THE WHEEL TO THE WORK SURFACE

11. Position the grinder table so that the wheel is over one corner of the surface to be ground.
12. Turn the wheelfeed handwheel to bring the grinding wheel close to the work surface.
13. Place a piece of paper between the work surface and lower the wheel until it just contacts the paper (Fig. 7-12).
14. Note the wheelfeed handwheel reading and reverse the handwheel about one-half turn to raise the wheel and then bring it back to within 0.005 in. of the original setting.
15. Traverse the workpiece under the wheel while rotating the wheel by hand to locate the high spot of the work surface (Fig. 7-13).
16. Lower the wheelfeed 0.001 in. for each pass across the work until the high spot is located.
17. Move the table so that the wheel clears the edge of the work surface to be ground.

Fig. 7-12 Using a piece of paper to set the CBN wheel close to the work surface. *(Courtesy of GE Superabrasives)*

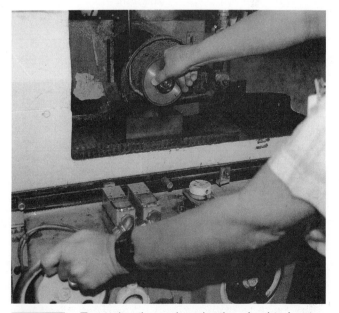

Fig. 7-13 Traversing the work under the wheel to locate the high spot of the work surface. *(Courtesy of GE Superabrasives)*

18. Use high-lubricity grinding fluids for efficient stock removal and long CBN wheel life.

19. Stop the grinder spindle and adjust the coolant nozzle so that it is about ¼ in. (6 mm) above the die surface and as close to the CBN wheel as possible.

20. Place a dummy block, slightly lower than the surface to be ground, at the right-hand end of the die so that the entire die surface receives coolant at all times (Fig. 7-14).

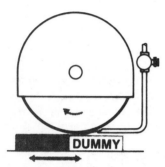

Fig. 7-14 A dummy block distributes coolant and ensures that the right-hand surface of the work is wet at the start of the cut. *(Courtesy of GE Superabrasives)*

GRINDING THE SURFACE

21. Start the grinder spindle and lower the wheelhead 0.001 in. (0.02 mm) for the first cut.

22. Start the coolant flow, ensuring that a good supply is directed to the point of wheel-workpiece contact.

23. Start the table reciprocating and engage the crossfeed to take a roughing pass across the die surface.

24. Be sure that the edge of the grinding wheel completely clears the side of the die after a pass over the die surface.

25. Take as many passes at 0.001 in. (0.02 mm) depth of cut in order to sharpen the die properly.

26. For the final pass, set the wheelhead for 0.0005 in. (0.01 mm) depth of cut to improve the surface finish.

27. Stop the table movement and then shut off the coolant.

28. Allow the wheel to revolve for about one-half

29. Shut off the grinder.

SIDEWHEEL AND SLOT GRINDING

Cubic boron nitride grinding wheels are especially well suited for sidewheel and slot grinding applications because of their consistent sharpness and low wear rates (Fig. 7-15). The following guidelines should be followed for best results.

Fig. 7-15 CBN wheels are excellent for sidewheel grinding operations. *(Courtesy of GE Superabrasives)*

1. *Undercut the Wheel Core.*

   —The wheel core side should be undercut about 0.010 in. (0.25 mm) to prevent it from rubbing against the workpiece (Fig. 7-16).

2. *True and Condition the Wheel.*

   —Mount the wheel securely to prevent any wobble.

   —True and condition the side of the wheel (see Chapter 5).

3. *Use Normal Wheel Speeds.*

   —Speeds of 5000 to 6500 sf/min. (25 to 33 m/s) are recommended for sidewheel and slot grinding.

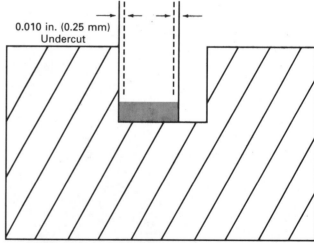

0.010 in. (0.25 mm) Undercut

0.0005–0.001 in. (0.01–0.02 mm) Crossfeed increments

Fig. 7-16 Undercutting the wheel core prevents it from rubbing against the workpiece during sidewheel grinding. *(Courtesy of GE Superabrasives)*

4. *Use Lower Table Speeds.*
—Table speeds should be somewhat slower than for conventional grinding.
—Use table speeds of 30 to 50 ft/min (0.15 to 0.25 m/s) for fast grinding to size.
—Use 12 to 20 ft/min (0.06–0.10 m/s) for finishing passes.

5. *Use Small Crossfeed Increments.*
—Crossfeed increments in the range of 0.0005 to 0.001 in. (0.01 to 0.02 mm) are recommended.

6. *Use Correct Coolant.*
—A good supply of 5 to 10 percent heavy-duty soluble oil coolant is recommended for best grinding wheel performance.

CREEP-FEED GRINDING

Creep-feed grinding is a technique used to grind a form into a workpiece in a single pass of the grinding wheel. The workpiece is fed into the revolving grinding wheel, opposite to the wheel rotation, at a slow, steady table feed rate. The wheel height is set to the final size and the desired form is generally completed to size, tolerance, and surface finish in one pass.

Creep-feed grinding can be competitive with surface grinding, milling, gear cutting, broaching, and other processes where heavy stock removal is required. It is very effective where a precise, accurate form is required, and profile accuracy is critical. The key advantages to creep-feed grinding are increased productivity, better dimensional part accuracy, and less metallurgical damage to the workpiece.

Creep-Feed Grinders

Because of the grinding pressures involved, true creep-feed grinding requires machines which have been specifically designed for the creep-feed grinding process. These machines must be rigid, and both statically and dynamically stable.

There are three basic types of creep-feed grinding:

1. *Pseudo-Creep-Feed grinding.* This is used for applications where the workpiece is narrow.
2. *True Creep-Feed grinding.* This generally requires machines which are specifically designed for creep-feed grinding.
3. *Continuous-Dress Creep-Feed (CDCF) grinding.* This requires special-purpose machines that also have compensating spindle speeds to maintain constant grinding wheel surface speeds. In general, CBN wheels are not used on this type of equipment.

Creep-Feed Grinding Process

Cubic boron nitride wheels are recommended for creep-feed grinding of hardened tool and alloy steels, hard cast irons, and nickel-base and cobalt-base superalloys. Because of the extremely low rate of wheel wear, wheel straightness and form on CBN wheels are normally maintained through hundreds of grinding cycles. They remove stock by an efficient chip-making process and self-sharpen continuously during grinding to ensure that high stock-removal rates are possible without dulling the wheel.

Creep-feed grinding differs from conventional surface grinding because the grinding wheel contact is fairly large, generally 1 to 2 in., and the metal-removal rates are high. The long arc of contact tends to reduce any machine vibrations. It also means that there are more abrasive grits in contact with the workpiece at any one time and that a significant amount of heat and chips are generated which require a good supply of coolant to carry away the chips and keep the grinding heat to a minimum. The table feed rate is dependent on the width and depth of the form being produced. A 1-in. or 25-mm per minute feed rate is fairly common for most creep-feed grinding applications.

Creep-Feed Grinding Guidelines

The following guidelines should be followed in order to obtain the best creep-feed grinding results.

- True and dress the CBN wheel properly before attempting any grinding.

- Hold the workpiece securely to prevent any vibration which could affect the grinding operation.

- Set the coolant supply nozzle(s) slightly above the work surface and as close as possible to the work-wheel interface. Two opposed high-pressure coolant nozzles are recommended for best results (Fig. 7-17).

Fig. 7-17 Creep-feed grinding is used to finish grind a form in one pass of the wheel. *(Courtesy of GE Superabrasives)*

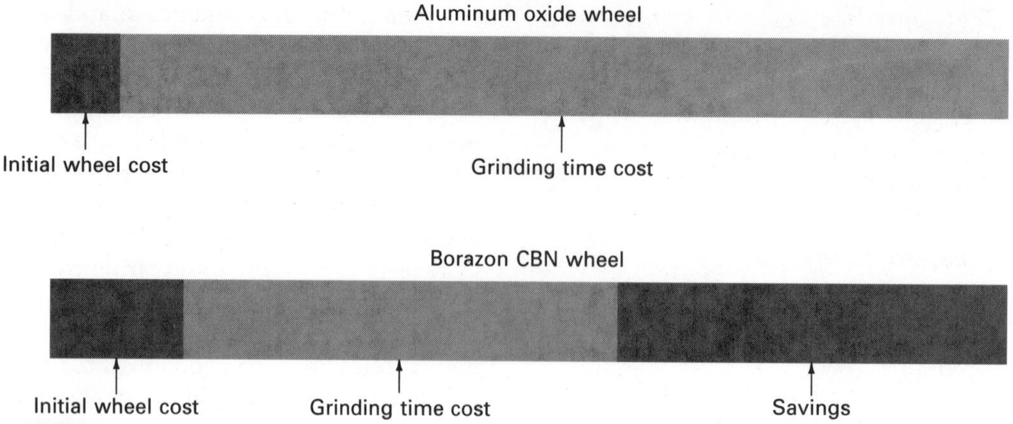

Aluminum oxide wheel

Initial wheel cost   Grinding time cost

Borazon CBN wheel

Initial wheel cost   Grinding time cost   Savings

Fig. 7-18 The cost-effectiveness of CBN grinding wheels. *(Courtesy of GE Superabrasives)*

- Position the workpiece in the correct relationship to the grinding wheel.
- Set the CBN wheel to the proper depth for the form required.
- Set the table feed rate to 1 in. (25 mm).

COST-EFFECTIVENESS OF CBN WHEELS

Cost-effective surface-grinding applications of CBN grinding wheels cover practically any carbon or alloy steel that has a Rockwell C 50 hardness or higher. This includes stainless and bearing steels. Extremely hard, abrasion-resistant cast irons and abrasion-resistant steel alloys can be efficiently ground with CBN wheels. The benefits of CBN wheels for surface-grinding applications are consistent part accuracy, higher productivity, and lower grinding costs.

The cost-effectiveness of CBN wheels is illustrated in Fig. 7-18. Both aluminum oxide and CBN wheels were used to grind hardened steel broach bodies, and the results were analyzed and compared.

1. *Wheel Cost.* The wheel cost for the CBN wheel is higher than that for the aluminum oxide grinding wheel.
2. *Grinding Time Cost.* Labor and overhead costs for the CBN wheel are lower as a result of:
   a. Reduced grinding time
   b. Fewer wheel changes
   c. Less wheel conditioning
   d. Fewer stops for gaging and machine adjustments to compensate for wheel wear
   e. Faster finishing cycles
3. *Total Grinding Cost per Tool.* The total grinding cost is reduced by 40 percent, making CBN wheels very cost-effective and an excellent investment.

Cost Comparisons

The high material-removal rates of CBN wheels result in increased productivity and reduced total grinding costs. Heavy feed rates can be used without burning the workpiece, and the parts ground with CBN wheels last longer because of better tool geometry and less metallurgical damage.

A few surface-grinding operations performed with an aluminum oxide wheel and a CBN wheel follow to compare various grinding factors.

DIE GRINDING. A large pierce and blanking die, made of AISI M-4 steel hardened to Rc 62, required regrinding (Fig. 7-19). To regrind this die, 0.800 in.^{3} (13.1 cm^3) of stock had to be removed. The advantages of the CBN wheel over the aluminum oxide wheel are as follows:

1. Grinding time was reduced from 3.5 h with an aluminum oxide wheel to 1 h with a CBN wheel.
2. Grinding costs, based on a labor and overhead (LOH) rate of $40.00 per hour, were reduced from $140.00 to $40.00.

Fig. 7-19 Regrinding a large pierce and blanking die with a CBN wheel reduced the grinding costs and improved die life. *(Courtesy of GE Superabrasives)*

3. The aluminum oxide wheel was dressed four times before the die was ground, while the CBN wheel required no dressing.
4. No extra grinding passes were required to compensate for wheel wear.
5. The die ground with the CBN wheel showed excellent life in the shop because there was no metallurgical damage.

SINGLE-POINT NUMERICALLY CONTROLLED TOOL GRINDING. The index faces of 346 single-point numerical control tools, made of AISI M-2 steel, had to be ground (Fig. 7-20). The amount of stock which had to be removed to complete this job was 40.47 in.^{3} (665 cm^3). The advantages of the CBN wheel over the aluminum oxide wheel are as follows:

1. Grinding time was reduced from 1 h for each tool to 40 min with a CBN wheel.
2. Grinding costs, based on a labor and overhead rate of $40.00 per hour, for the 346 tools were reduced from $13,840.00 to $9200.00.
3. Tools ground with the CBN wheel performed longer than did those ground with the aluminum oxide wheel because there was no metallurgical damage.
4. One aluminum oxide wheel was consumed for every 40 pieces ground. Therefore, it took 8.65 grinding wheels to grind the 346 pieces, while it took only one CBN grinding wheel to grind the same number.

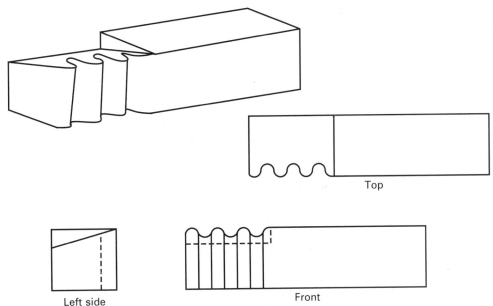

Fig. 7-20 Single-point numerical-control cutting tools can be effectively ground with CBN wheels. *(Courtesy of GE Superabrasives)*

REVIEW QUESTIONS

Surface Grinding

1. List three advantages of using CBN wheels for surface grinding.
2. What are the advantages of the cool-cutting CBN grinding wheels?

Types of Surface Grinders

3. Name the four basic types of surface grinders.

Types of Material Ground

4. List the three types of steels which are most effectively ground with CBN wheels.

5. Define grindability.

Preparing the Grinder

6. List four of the most important key points in the use of CBN wheels for surface grinding.

Surface Grinding with CBN Wheels

7. What CBN wheel is recommended for most standard surface-grinding operations?
8. What type of coolant is recommended for most surface-grinding operations?
9. State the recommended speeds and feeds for surface grinding.

Sidewheel and Slot Grinding

10. List the guidelines which should be followed when using CBN wheels for sidewheel and slot grinding.

Creep-Feed Grinding

11. What is creep-feed grinding?

12. What characteristics do creep-feed grinders possess?

13. List the six guidelines which should be observed for creep-feed grinding.

Cost-Effectiveness of CBN Wheels

14. Why is grinding time cost with CBN wheels much less than with conventional wheels?

Cylindrical Grinding

The development of hardened steel in the latter part of the nineteenth century created a need for a machine which was capable of finishing workpieces which were as hard as cutting tools. This lead to the development of grinders, which over the years were improved and modified to become the high-precision grinders of today. *Cylindrical grinding* involves the grinding of the periphery of a rigidly held rotating workpiece (Fig. 7-21). Although there are many forms of cylindrical grinding, this section will deal with only the center-type cylindrical grinder.

OBJECTIVES

After completing this section you should be able to:

1. Select the proper CBN grinding wheel to suit the material to be ground
2. Mount, true, and dress the CBN wheel
3. Grind work on a center-type cylindrical grinder

CUBIC BORON NITRIDE

The development of CBN abrasive provided industry with grinding wheels that were much harder, would last much longer, would remove material faster, and would produce better-quality products than conventional grinding wheels.

- The first-generation CBN superabrasives, called *monocrystals* (Fig. 7-22A), were designed for the efficient and cost-effective grinding of hardened tool, die, and alloy steels, and also high-temperature superalloys with a hardness of Rc 35 and higher.

- The second-generation CBN superabrasives, called *microcrystalline CBN superabrasives* (Fig. 7-22B), perform exceptionally well when grinding some hardened alloy steels and tough superalloys. They were designed primarily for the efficient, cost-effective grinding of easier-to-grind (ETG) ferrous materials such as mild steels, medium-hard alloy steels, stainless steels, cast irons, forged steels, and some superalloys.

MACHINE SPECIFICATIONS

Modern abrasives (superabrasives) and machines have developed the art of grinding into a highly effi-

Fig. 7-21 CBN superabrasive wheels are finding wide use in cylindrical grinding operations. *(Courtesy of GE Superabrasives)*

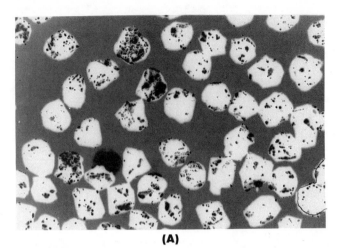

(A)

Fig. 7-22A Monocrystalline CBN is blocky in shape and consists of large crystals. *(Courtesy of GE Superabrasives)*

cient production tool. Work tolerances on a production basis have gradually been reduced from one thousandth of an inch to within a few millionths of an inch. *Surface quality*, which in the early 1900s was defined in terms such as "rough," "commercial," and "high," is now consistently measured and produced in microinches. All of this has been possible because of the new abrasive products which were developed and the special grinders which were designed for superabrasive wheels to produce accurate parts and reduce grinding costs (Fig. 7-23).

Machines used to grind with CBN wheels should have the following characteristics:

- High rigidity, freedom from chatter and vibration

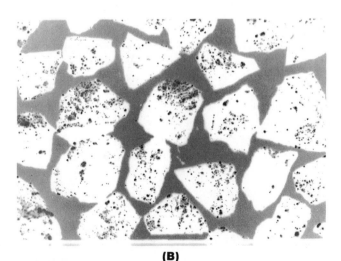

(B)

Fig. 7-22B Microcrystalline CBN is irregular in shape and consists of micrometer-size crystals strongly bonded to each other to form a supertough superabrasive. *(Courtesy of GE Superabrasives)*

- Very high spindle speeds
- Accurate programmable feed systems
- Wet-grinding capabilities
- Special wheel truing and dressing systems

When existing machines have, or can be modernized to have, these characteristics, CBN wheels will usually perform well and be cost-effective. The effects of worn spindle bearings, loose slides, sloppy feed screws, and nonresponsive feed systems cannot be overcome simply by installing a CBN wheel; nor can a grinder designed to use conventional grinding wheels be expected to fully utilize the full productive potential of CBN wheels.

Machines designed to grind with CBN wheels can be five to ten times more productive than machines designed to grind with conventional grinding wheels. Because of their high productivity and the quality of work produced on specially designed grinders, they often provide a very attractive return on investment (ROI).

See Chapter 5 for a detailed explanation of the qualities that a grinder should have to:

- Increase productivity
- Reduce machine downtime
- Reduce production costs
- Produce high-quality workpieces
- Produce consistent surface finishes

WHEEL SELECTION

The characteristics of superabrasive grinding wheels must suit the workpiece material and the machine on which it is to be used if optimum grinding efficiency is to be achieved. CBN grinding wheels have been designed to:

1. Reduce working pressures at the wheel-work interface for greater dimensional accuracy.
2. Provide rapid material removal rates for higher production.
3. Improve surface quality in parts for greater wear resistance.

Resin-bond wheels made with CBN superabrasives provide a cost-effective combination of long wheel life and high material-removal rates. Grinding ratios (the ratio of the volume of work material removed to the volume of wheel material used during grinding) are generally 100 times higher than those for grinding the same materials under the same conditions with conventional abrasives. One resin-bond CBN wheel generally outlasts dozens of aluminum oxide wheels.

When selecting a CBN wheel for a cylindrical

Cylindrical grinders using CBN wheels should be of rigid construction and have good spindle bearings to utilize the full potential superabrasive wheels. *(Courtesy of Cincinnati Milacron Co.)*

grinding operation, the following points should be considered:

1. *Grit Size.* The size of the abrasive grain affects the material-removal rate and the type of surface finish which will be produced.
   a. It is best to choose the largest superabrasive grit size that produces the desired finish because it usually has a longer life.
   b. A wheel with a small grit size may produce smoother surface finishes if feed rates are lowered.
2. *Concentration.* The material-removal rates and the life of CBN wheels are strongly influenced by the superabrasive concentration (amount of superabrasive per unit volume of wheel bond material). Wheels with high (75 to 100) superabrasive concentration are usually the most cost-effective because they last longer (Fig. 7-24).
3. *Wheel Size.* When a CBN wheel replaces a conventional grinding wheel, it should be of the same diameter and width as the wheel it replaces, provided the grinder is rigid and has the power required for the superabrasive wheel.

If it is necessary to use an undersize wheel, reduce the width of the wheel but not the diameter because surface speed is very important to the performance of CBN wheels.

Wheel Selection Guidelines

The following general selection criteria should be used as a guide in selecting CBN wheels for cylindrical grinding operation. Since there are so many variables involved in a grinding operation, these guidelines may have to be varied somewhat to suit the particular machine tool, workpiece material, and grinding conditions.

1. *Resin-bond wheels* containing 120-grit-size abrasive at 100 concentration can generally be used for all startup cylindrical grinding operations with CBN abrasive.
2. *Grit size* may be varied as required to suit the surface finish required.

   —Wheels containing finer-grit-size abrasive produce good surface finishes and are generally freer cutting and easier to true and dress.
3. *Higher concentrations* of CBN provide longer wheel life and generally require more power to grind. These wheels are also more difficult to true and dress.
4. *Vitrified bond wheels* are generally used in grinding operations where a high degree of wear resistance is required.

   —Because the vitrified bond is much stronger than the resin bond, higher concentrations of CBN can be used without excess pullout.

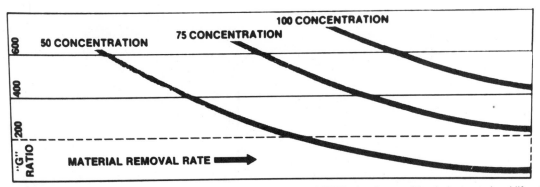

Fig. 7-24 High concentrations improve the performance of CBN wheels, resulting in longer wheel life at higher material-removal rates. *(Courtesy of GE Superabrasives)*

—Concentrations of 150 and 200 (38 and 50 vol % CBN) are common for vitrified wheels.

5. *Electroplated wheels* are generally used in form grinding operations where precise control of the shape generated by the wheel is required.

NOTE Be sure to read Chapter 4, which gives a detailed explanation of wheel selection, before using any CBN wheel. CBN wheel manufacturers test wheels extensively, and their recommendations for use should be closely followed.

WHEEL MOUNTING AND PREPARATION

Chapter 5 gives a detailed explanation of mounting and preparing a CBN wheel for grinding. Be sure to refer to that chapter before proceeding with this operation. The following are intended only as a reminder of some of the key points which must be covered if the grinding operation is to be successful.

1. For best results with CBN wheels, a high-quality wheel adaptor should be used, and wherever possible the wheel and adapter should stay together as a unit for the life of the grinding wheel.
2. Use a dial indicator to true the wheel within 0.001 in. (0.02 mm) or less runout on the wheel circumference (Fig. 7-25).
3. To minimize chatter, vibration, and wheel wear, CBN wheels should be balanced after mounting. Dynamic balancing is recommended for large CBN wheels.
4. Use the appropriate truing device for the type of CBN bond.
5. Dress the wheel to expose sharp CBN abrasive cutting edges.

NOTE Electroplated bond CBN wheels do not require truing or dressing.

WORK PREPARATION AND MOUNTING

To produce accurate work, good surface finishes, and cost-efficient material removal rates, it is important

that the workpiece be prepared and mounted properly. Let us assume that the diameter of a shaft will be ground with the workpiece mounted between centers. Several basic factors apply to all cylindrical grinding operations and should be followed for the best grinding conditions.

1. Align the centers of the grinder with a test bar and dial indicator to ensure that a parallel diameter will be produced.
2. Hone or lap the center holes of the workpiece to provide accurate work rotation.
3. Adjust the headstock and footstock centers to fit the length of the workpiece. Set them at approximately equal distances from the table center.
4. Apply a suitable lubricant to the center holes of the work.
5. Place a suitable driving dog on the work and then mount the workpiece between centers.
6. Adjust the footstock center tension so that the

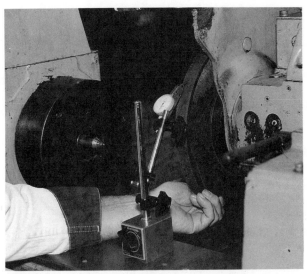

Fig. 7-25 Using a dial indicator to true a CBN wheel to within 0.001-in. (0.02-mm) runout on the circumference. *(Courtesy of M. Rapisarda and GE Superabrasives)*

work is held firmly between centers. Avoid excess pressure.

7. Support long, slender workpieces with a suitable number of work rests to prevent springing. The distance between each work rest should be approximately six to ten times the diameter of the work (Fig. 7-26).

SPEEDS, FEEDS, AND DEPTH OF CUT

Many factors govern the efficiency of a cylindrical grinding operation. Some of the key factors are grinding-wheel speed, workpiece speed, depth of cut, and traverse feed rate (Table 7-4). These factors affect the life of the CBN wheel, the material-removal rate, the type of surface finish produced, and the accuracy of the workpiece.

Wheel Speed. CBN wheels perform best at high surface speeds. Increasing speeds result in higher material removal rates and longer wheel life (Fig. 7-27). The normal range of speeds for CBN wheels is 5000 to 6500 sf/min; however, some grinders are capable of speeds of 20,000 sf/min (102 m/s) and higher. *In no case should the recommendations of the grinding-wheel manufacturer be exceeded.*

Work Speed. The workpiece must rotate opposite to the grinding-wheel rotation and be correlated with the wheel speed and abrasive wheel characteristics to obtain the best material-removal rates. For CBN cylindrical grinding, the work speed gen-

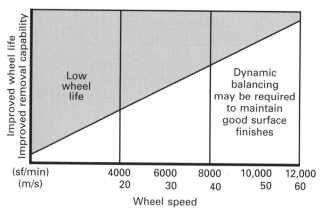

Fig. 7-27 CBN grinding wheels perform best at high surface speeds. *(Courtesy of GE Superabrasives)*

erally ranges from 50 to 200 sf/min. The best surface finishes are produced at 50 sf/min, while the CBN works harder at 200 sf/min and removes metal faster.

Depth of Cut. The recommended depth of cuts for CBN cylindrical grinding, depending on work diameter, range from:

- 0.0005 to 0.002 in. (0.01 to 0.05 mm) per pass for rough grinding operations.

- 0.0002 to 0.0005 in. (0.005 to 0.01 mm) per pass for finish grinding operations.

Fig. 7-26 Long, slender workpieces should be supported by a suitable number of work rests to prevent springing during the grinding operation. *(Courtesy of Cincinnati Milacron Co.)*

Table 7-4 CBN CYLINDRICAL GRINDING SPECIFICATIONS

Material	Condition	Wheel Speed		Work Speed		Diameter Infeed (Wheel Width/rev)		Traverse Rate (Wheel Width/rev)	Wheel Identification (ANSI)
		ft/min	m/s	ft/min	m/min	in.	mm		
Carbon steels (low, medium, high carbon)									
Free machining carbon steels (low, medium carbon)	>Rc 50								
Carbon steels, cast (low, medium carbon)									
Free machining alloy steels (leaded, resulfurized)	Carburized and/or quenched and tempered	5000–7500	25–38	50–100	15–30	0.001 (rough) 0.0002 (finish)	0.025 (rough) 0.005 (finish)	¼ (rough) ⅛ (finish)	B100T 100B
Alloy steels (low, medium, high carbon)									
Alloy steels, cast (low, medium carbon)									
Tool steels, cast									
High-strength steels	>Rc 50			50–100		0.001 (rough) 0.0002 (finish)	0.025 (rough) 0.005 (finish)	¼ (rough) ⅛ (finish)	B100T 100B
Tool steels, wrought					15–30				
High-strength steels, wrought	Quenched and tempered	5000–7500	25–38		15–30				
Nitriding steels, wrought	Rc 60–65 nitrided	5000–7500	25–38	50–100	15–30	0.001 (rough) 0.0002 (finish)	0.025 (rough) 0.005 (finish)	¼ (rough) ⅛ (finish)	B100T 100B
Maraging steels, wrought	>Rc 50 maraged	5000–7500	25–38	50–100	15–30	0.001 (rough) 0.0002 (finish)	0.025 (rough) 0.005 (finish)	¼ (rough) ⅛ (finish)	B100T 100B

Traverse Feed Rate. The traverse feed rate affects the material removal rate and the surface finish produced. For CBN cylindrical grinding, the following traverse feed rates are recommended:

- One-quarter to one-third the width of the CBN wheel face per workpiece revolution for general-purpose cylindrical grinding.

- One-eighth the width of the CBN wheel face per workpiece revolution for finish grinding operations.

GRINDING A PARALLEL DIAMETER WITH A CBN WHEEL

Cubic boron nitride wheels were designed to be worked hard and remove material as quickly as possible. Grinding conditions should be established for optimum productivity rather than for the longest possible wheel life. This generally results in the lowest possible total grinding cost per part. If the proper wheel has been selected and conditioned for the job, the machine is in good condition and has adequate horsepower, CBN wheels will provide great opportunities for improving productivity.

Procedure

1. Mount, true, and dress the proper CBN wheel for the grinding operation.
2. Align the headstock and footstock centers to ensure a parallel diameter.
3. Mount the workpiece between the centers.
4. Start the grinding wheel and allow it to run for a few minutes to warm up the spindle bearings.

5. Adjust the table reversing dogs to allow some overtravel at each end of the workpiece (Fig. 7-28).

NOTE The left-hand reverse dog must be set so that the table reverses slightly before contacting the driving dog or work shoulder.

6. Set the grinder to the proper speed for the size and type of CBN wheel being used.
7. Set the work speed to suit the diameter and type of work material being ground.
8. Move the table so that the full wheel width is over the unground workpiece.
9. Carefully plunge feed the wheel into the work diameter until a concentric diameter is produced.
10. Stop the grinding wheel and workpiece rotation.
11. Measure the diameter of the ground (concentric) section and calculate the amount yet to be removed.
12. Set the depth of the feed index so that the wheel stops approximately 0.002 to 0.003 in. (0.05 to 0.07 mm) before the finish diameter of the workpiece.
13. Set the machine feed rates for roughing.
14. Start the wheel and work rotation and move the table so that the full wheel width is over the work.
15. Start the coolant flow so that a good supply is directed at the work-wheel interface (Fig. 7-29).
16. Engage the automatic feed mechanism and plunge grind to the rough cut size.
17. Withdraw the wheel from the work and move the table so that the full wheel width is over the next section of the work.

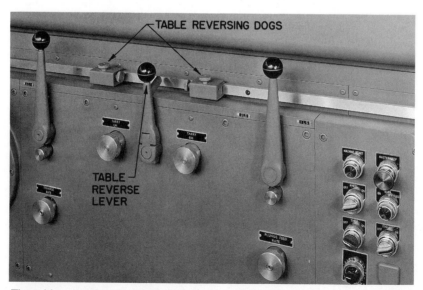

Fig. 7-28 The table-reversing dogs should be set to suit the length of the section to be ground and so that only one-third of the wheel face overlaps the end of the work. *(Courtesy of Landis Tool Co.)*

18. Repeat operations 16 and 17 and plunge grind the complete section to be ground to the roughing diameter.
19. Set the machine controls for traverse grinding.
20. Move the wheel to within 0.001 or 0.002 in. (0.02 to 0.05 mm) of the workpiece rough diameter.
21. With the coolant on and the infeed off, engage the table traverse.
22. Set the table traverse feed to one-eighth the width of the CBN wheel face per workpiece revolution.
23. Take finish cuts of 0.0002 to 0.0005 in. (0.005 to 0.01 mm) until the diameter is ground to size.

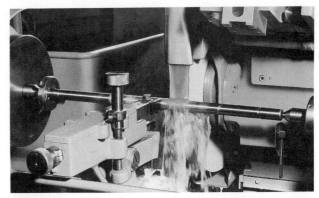

Fig. 7-29   Apply a good supply of coolant to the work-wheel interface for the best material-removal rates. *(Courtesy of Cincinnati Milacron Co.)*

REVIEW QUESTIONS

Cubic Boron Nitride Wheels

1. Name four advantages which the development of CBN wheels has provided industry.
2. Briefly describe the two generations of CBN superabrasives and state where each is used.

Machine Specifications

3. List five characteristics which grinders using CBN wheels should have.

Wheel Selection

4. What type of CBN wheel is generally used for cylindrical grinding operations?
5. Name the three factors which should be considered when selecting a CBN wheel.

Wheel Mounting and Preparation

6. List five key factors which must be considered when mounting and preparing a CBN wheel.

Work Preparation and Mounting

7. List four of the most important factors which apply to mounting a workpiece for cylindrical grinding.

Speeds, Feeds, and Depth of Cut

8. What is the normal range of speeds for CBN wheels?
9. Explain the effects of increasing the wheel speed.
10. What work speeds are generally used when cylindrical grinding with CBN wheels?
11. List the recommended depth of cuts for cylindrical grinding.
12. What traverse feedrate is recommended for cylindrical grinding?

Internal Grinding

Since the commercial introduction of CBN by GE in 1969, the use of this abrasive has grown at a steadily expanding rate. After considerable research and experimentation, it was found that CBN wheels could be used successfully for internal-grinding operations. The advantages which CBN abrasives bring to other forms of grinding also applied to internal-grinding operations. These were:

• High stock removal

• Long wheel life

• Greater precision

• Higher-quality components

After completing this section you should be able to:

1. Select the proper CBN wheel for the type of material to be ground
2. Mount, true, and dress the wheel for the best performance
3. Set speeds and feeds for internal-grinding operations

MACHINE SPECIFICATIONS

Modern abrasives (superabrasives) and machines have developed the art of grinding into a highly efficient production tool. Work tolerances on a production basis have gradually been reduced from one thousandth of an inch to within a few millionths of an inch. Surface quality is now consistently measured and produced in microinches. All this has been possible because of the new abrasive products which were developed and the special grinders which were designed for superabrasive wheels to produce accurate parts and reduce grinding costs (Fig. 7-30).

Machines used to grind with CBN wheels should have the following characteristics:

- High rigidity and freedom from chatter and vibration
- Very high spindle speeds
- Accurate programmable feed systems
- Wet-grinding capabilities
- Special wheel truing and dressing systems

A rigid machine in good condition is required to obtain the best results from CBN wheels in internal-grinding operations. Machines designed to grind with CBN wheels can be five to ten times more productive than machines designed to grind with conventional grinding wheels. Because of their high productivity and the quality of work produced on specially designed grinders, they often provide a very attractive return on investment (ROI).

See Chapter 5 for a detailed explanation of the qualities that a grinder should have to:

- Increase productivity
- Reduce machine downtime
- Produce high-quality workpieces
- Produce consistent surface finishes

WHEEL SELECTION

Any successful grinding operation depends to a large extent to choosing the right wheel for the job (Fig. 7-31). The type of wheel selected and how it will be used will affect the material-removal rate and the life

of the grinding wheel. The selection of a CBN grinding wheel can be a complex task and it is always wise to follow the manufacturer's suggestion for each type of wheel.

Four factors generally must be considered when selecting a CBN wheel for an internal grinding operation:

1. *Bond.* Three types of wheel bonds are used for internal-grinding wheels.
   a. *Resin-bond wheels* contain a high concentration (100) of CBN abrasive. They are used for most

Fig. 7-30 Internal grinders using CBN wheels should be of rigid construction and have good spindle bearings for the best results. *(Courtesy of GE Superabrasives)*

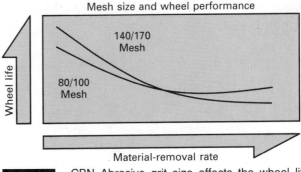

CBN Abrasive grit size affects the wheel life and material-removal rate. *(Courtesy of GE Superabrasives)*

A variety of CBN wheel sizes and styles are available for internal-grinding operations. *(Courtesy of GE Superabrasives)*

internal-grinding operations, giving an optimum combination of fast material removal and long wheel life.

b. *Vitrified-bond wheels* give good performance on all types of steels. They generally last longer than do resin-bond wheels and require little or no dressing. They require special handling to prevent damage.

c. *Electroplated wheels* provide excellent internal grinding results when using small CBN wheels and pins. Pins are mounted grinding wheels, generally 3/8 in. (9.5 mm) or smaller, which are driven by a steel shank. These wheels provide high material-removal rates but because they have a single layer of abrasive around the steel shank, they produce a rougher surface finish and do not last as long as resin-bond or vitrified-bond wheels or pins.

2. *Grit Size.* CBN abrasive grit size greatly influences material-removal rates and surface finishes. For high material-removal rates, use coarse-grit-size wheels; for good surface finishes, use fine abrasive grit sizes. When a considerable amount of stock must be removed from a workpiece and a fine surface finish is required, it is recommended to:

a. Rough out the hole with a coarse abrasive CBN wheel

b. Change to a wheel containing fine abrasive for the finishing operation

A good compromise is to use a wheel with 150-grit-size abrasive which provides an efficient material removal rate and produces a surface finish in the 15- to 20-μin. (AA) range (Fig. 7-32).

3. *Concentration.* The material-removal rate and the life of CBN wheels are strongly influenced by the superabrasive concentration in the grinding wheel. Wheels having 100 superabrasive concentration are usually the most cost-effective because they last longer.

4. *Wheel Size.* The best wheel size for an internal-grinding operation is one with a diameter of approximately 70 to 80 percent of the hole size.

a. Always select as large a wheel as possible so that more abrasive crystals can share the work, and thus cut faster, last longer, and be most cost-effective.

b. An exception to this general rule is for holes 3/4 in. (19 mm) or smaller, where wheel diameters should be less than 70 percent to leave enough room for good coolant flow.

c. The length of an internal CBN wheel should be 1/16 in. (1.5 mm) longer than the bore to be ground, if possible.

NOTE Be sure to read Chapter 4, which gives a detailed explanation of wheel selection, before using any CBN wheel. CBN wheel manufacturers test wheels extensively, and their recommendations should be closely followed.

WHEEL MOUNTING AND PREPARATION

Chapter 5 gives a detailed explanation of mounting and preparing a CBN wheel for grinding. Be sure to refer to this section before proceeding with this operation. The following apply especially to internal-grinding wheels and should be followed closely:

1. The spindle quill or arbor on which the CBN wheel is mounted should have as large a diameter as possible.

2. Keep the wheel overhang to a minimum to minimize deflection, chatter, and vibration (Fig. 7-33).

3. Mount the wheel securely on the spindle.

4. Use a dial indicator to true the wheel within

Fig. 7-33 Wheel overhang should be kept to a minimum to reduce chatter and vibration when internal-grinding. *(Courtesy of GE Superabrasives)*

0.001 in. (0.02 mm) or less runout on the wheel circumference.

5. Use the appropriate truing device for the type of CBN wheel bond (Fig. 7-34).
6. Dress the wheel to expose sharp CBN abrasive cutting edges.

NOTE Electroplated CBN wheels or pins do not require truing or dressing.

SPEEDS AND FEEDS

Cubic boron nitride wheel speed and feed rate will affect the material-removal rate and the life of the internal grinding wheel. The following general guidelines should give excellent results when internal grinding with a CBN wheel. These guidelines should be varied somewhat so that the best material-removal

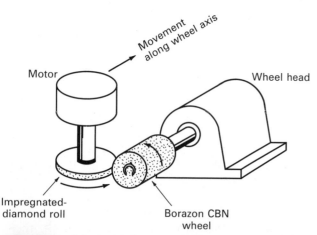

Fig. 7-34 A typical truing arrangement for internal-grinding wheels. *(Courtesy of GE Superabrasives)*

rate and wheel life are achieved for each internal-grinding operation. (Table 7-5)

1. *Wheel Speed.* Most steels can be internal-ground effectively at conventional speeds. Some applications, however, may require higher wheel speeds. The following are some CBN wheel speed recommendations.
   a. 10,000 to 12,000 sf/min (50 to 62 m/s) are recommended for the best performance of CBN internal-grinding wheels especially for small (<1 in) bores.
   b. 4000 to 6000 sf/min (20 to 30 m/s) are more generally used to grind holes greater than 1 in diameter.
   Higher wheel speeds generally provide higher material-removal rates, better CBN wheel life, and superior surface finishes.
2. *Feed Rate.* The traverse feed rate affects the material-removal rate and the surface finish produced. For CBN internal grinding, the following feed rates are recommended.
   a. One-quarter to one-third the width of the CBN wheel face per workpiece revolution for general-purpose internal grinding.
   b. One-eighth the width of the CBN wheel face per workpiece revolution for finish grinding operations.

OPERATING PROCEDURES

Probably the most important thing that will keep CBN wheels performing at high levels is the proper use of grinding fluid. Grinding fluid (coolant) serves five important functions:

- It removes heat from the wheel and workpiece.
- It lubricates the grinding operation reducing friction at the wheel–work interface.
- It keeps the wheel clean and prevents it from loading up.
- It washes chips and debris out of the internal diameter.
- It helps to retard oxidation of freshly ground surfaces.

The best coolant (fluid) to use for internal grinding operations is a 5 to 10 percent heavy-duty soluble oil. It should be applied as a heavy flood to the work–wheel interface to flush out chips effectively (Fig. 7-35).

In-Process Truing and Dressing

The frequency at which an internal-grinding wheel requires truing and dressing depends on the hole profile and taper requirements. Because of the slow,

Table 7-5   CBN INTERNAL-GRINDING CHECKLIST

Condition	Possible Cause	Corrective Action
1. Poor finishes	Chatter, vibration	—Make sure the wheel is not slipping on the mount —True the wheel
	Excessive dressing	—Use lighter dressing pressure —Stop dressing as soon as wheel starts to consume stick rapidly
	Abrasive grit size; too coarse	—Use finer grit size
	Poor coolant flow or location	—Apply heavy flood of 5–10 percent heavy-duty soluble oil so it reaches wheel–work interface
2. Part taper	Poor wheel alignment	—Correct wheel force deflection by rotating workhead 1 to 3°
	Poor truing	—Retrue
	Nonuniform coolant flow	—Adjust spout location for maximum coolant circulation
Corner cutting	Unequal reciprocation	—Equalize reciprocation
3. Bellmouth	Stroke too long	—Reduce stroke length
Barreling	Stroke too short	—Increase stroke length
4. Wheel won't cut	Glazed by truing	—Dress lightly until wheel opens up
	Wheel loaded	—Dress lightly until wheel opens up —Increase coolant flow to keep wheel surface clean —Never run wheel with coolant turned off (as common with aluminum oxide during wheel setup)
5. Slow cutting	Low feeds and speeds	—Increase feed rate —Increase wheel speed to 10,000–12,000 sf/min (50–60 m/s)
6. Short wheel life	Incorrect coolant flow	—Apply coolant to flood wheel–work interface
	Incorrect grinding fluid	—Use 5–10 percent heavy-duty soluble oil
	Excessive truing	—Don't try to remove all crayon marks —After initial truing, grind 100–200 pieces, true 0.0002-in. (0.005-mm) infeed increment, and continue grinding
	Low wheel speed	—Increase wheel speed
	Excessive dressing	—Dress only until wheel starts to consume stick —Do not dress after light truing; wheel will usually open itself up under correct grinding conditions

even wear rate of CBN wheels, a light truing after grinding 100 to 200 parts may be all that is required. A truing unit which is capable of 0.0002 in. (0.005 mm)

A heavy flood of coolant should be applied to the work-wheel interface when internal-grinding. (Courtesy of Cincinnati Milacron Co.)

infeed increments is recommended. Some internal grinders are equipped with such a truing unit which adjusts machine settings to automatically compensate for the amount removed from the CBN wheel.

HIGH-PRODUCTION GRINDING

It was only natural that CBN wheels would find applications in the production grinding of internal surfaces. This type of grinding was typified by the frequent dressing of conventional abrasive wheels, which, in turn, reduced the life of the grinding wheel. It also required that production stop during the dressing operation and also the time required to change grinding wheels, which reduced the productive efficiency of the grinding machine.

Cubic boron nitride wheels are ideal for high-production internal-grinding systems because they offer long wheel life, durability, and high material-removal rates. Although CBN wheels are used for a wide variety of internal-grinding applications, their major use is in the bearing and automotive industries (Fig. 7-36). An extensive testing program of the use of

CBN wheels for internal-grinding applications leads to the following conclusions:

1. If properly applied in a suitable grinding system, CBN wheels can not only compete with but also offer significant cost advantages over conventional aluminum oxide wheels in high-production-rate internal grinding.
2. The grinding ratio achieved by CBN wheels at high material-removal rates may be extremely high (in the order of 1000 to 5000).
3. CBN wheels perform best at high surface speeds [8000 to 12,000 sf/min (40 to 60 m/s)].
4. CBN wheels provide optimum performance as the wheel diameter approaches 75 percent of the bore diameter.
5. The 140/170 grit size of Borazon CBN Type 1 at 25 vol % (100 concentration) was found to be very effective in vitreous-bonded wheels.
6. The maximum material-removal rate and the highest G ratios were obtained when using straight oil grinding fluids.
7. In some instances, material-removal rates may have to be compromised somewhat to obtain the proper surface finish on the workpiece.
8. An *in-process wheel truing and dressing system* is essential to realizing maximum production rates.

ECONOMICS OF CBN INTERNAL GRINDING

Cubic boron nitride internal grinding wheels can reduce the cost of grinding steels and cast irons with a hardness of Rc 50 and above and tough cobalt-base superalloys, Rc 35 and above. CBN abrasive wheels offer the following advantages over conventional abrasive wheels:

1. *Harder, Stronger, Tougher Abrasive*
   a. CBN abrasive is much harder and tougher than aluminum oxide abrasive and thus outlasts it by a wide margin.
   b. There is less wheel consumption per ground part, resulting in less downtime for wheel changes.
   c. Reduced wheel wear results in fewer production stops to make dimensional checks and adjustments.
2. *Grind Faster, Grind Cooler*
   a. Productivity is improved tenfold or more because more stock can be removed in fewer passes.

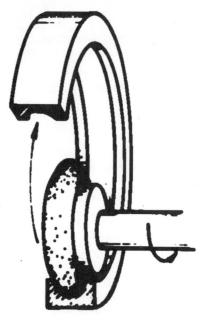

Fig. 7-36 The ball track of an outer bearing race being ground to size and shape with a CBN internal-grinding wheel. *(Courtesy of GE Superabrasives)*

   b. There is less chance of subsurface metallurgical damage to the part because of the cool cutting action.
   c. There is less chance of setting up undesirable residual stresses; therefore, there is less chance of part distortion and this results in longer part life.
3. *More Pieces per Wheel*
   a. At extremely high material-removal rates, CBN wheels last much longer than aluminum oxide wheels.
   b. CBN wheels stay sharp and free-cutting; therefore, the need for time-consuming wheel maintenance is reduced.
4. *Lowest Cost per Piece*
   a. CBN wheels produce a better quality part in less time.
   b. The net result is lower total grinding costs, which means lower cost per ground part.

Refer to Chapter 6 for a detailed explanation of the economics of using CBN wheels for grinding operations.

REVIEW QUESTIONS

Machine Specifications

1. List four advantages of CBN abrasives in the grinding process.

2. What characteristics should a grinder have to use CBN wheels effectively?

Wheel Selection

3. Name three types of bonds used for internal-grinding wheels.
4. How does grit size affect the grinding operation?
5. What grit size is recommended for efficient internal grinding?
6. What is the best size wheel to use for internal grinding?

Wheel Mounting and Preparation

7. List three important factors which should be considered when mounting an internal-grinding wheel.

Speeds and Feeds

8. What effect does the wheel speed and traverse feed rate have on the internal-grinding operation?
9. List three advantages of using higher wheel speeds for internal grinding.
10. What feed rate is recommended for internal grinding?

Operating Procedures

11. State five functions of a grinding fluid or coolant.

High-Production Grinding

12. Why are CBN wheels ideal for high-production internal-grinding systems?
13. What CBN wheel size provides optimum grinding performance?
14. What type of grinding fluid gives the maximum material-removal rate and highest G ratio?

Economics of CBN Internal Grinding

15. List four reasons why CBN abrasive is superior to conventional abrasive for internal-grinding operations.

CBN Internal Grinding Checklist

16. List the possible causes for the following internal-grinding conditions:
    a. poor finishes
    b. wheel will not cut

Tool and Cutter Grinding

Cubic boron nitride grinding wheels were first tested on difficult-to-grind (DTG) hardened tool and die steel cutting tools (Fig. 7-37). These steels are so hard and abrasion-resistant that they cause rapid dulling of the conventional aluminum oxide abrasive. As the abrasive dulls, the wheels tend to burnish rather than cut the workpiece material, thus reducing grinding productivity and causing metallurgical damage to the tool being ground.

In contrast, CBN wheels remain sharp and cool-cutting when grinding hardened tool and die steels. Because of their cool-cutting action, they cause little or no metallurgical damage; therefore, cutting tools ground with CBN wheels stay sharp longer than do those ground with aluminum oxide wheels.

OBJECTIVES

After completing this section you should be able to:

1. Select the proper CBN grinding wheel to suit the material to be ground
2. Prepare the grinder and the CBN wheel for grinding
3. Compare the advantages and cost-effectiveness of CBN wheels

CBN VERSUS ALUMINUM OXIDE ABRASIVE

Cubic boron nitride free-cutting grinding wheels remove stock quickly with little pressure and produce sharp edges on cutting tools with little or no thermal damage and better-than-new concentricity and tolerances. These super results are possible because CBN is a superabrasive that is second only to diamond in hardness. Some important factors which make CBN abrasive better than aluminum oxide for sharpening cutting tools made of hardened steels such as M-2, M-4, M-15, T-1, T-9, T-15, A-7, and D-7 are:

1. CBN abrasive retains its strength to above 1832°F (1000°C) and does not chemically react with steel.
2. CBN has twice the hardness and four times the

Fig. 7-37 CBN wheels were first tested on tool-and-cutter grinder for sharpening cutting tools. *(Courtesy of GE Superabrasives)*

Grinding Machine

The performance of a CBN wheel will vary with the condition of the grinder. A superabrasive grinding wheel will not work properly if at all on a grinder in poor condition (Fig. 7-38). For best results when grinding with CBN wheels, the grinder must have the following characteristics:

1. *Tight spindle bearings* to eliminate vibration and chatter. This lengthens the life of the grinding wheel, resulting in excellent surface finish and dimensional accuracy of the cutting tool.
2. *Close-fitting slides* to prevent chatter, which, in turn, reduces wheel effectiveness.
3. *Constant spindle speed* to boost efficiency. Variations in spindle speed reduce cutting efficiency and shorten wheel life. Constant spindle speeds along with high torque capacity are necessary for high metal-removal rates.

abrasion resistance of the aluminum oxide abrasive used in grinding wheels.

3. This combination makes CBN wheels especially well suited for the grinding of high-speed and superalloy steel tools. They provide:
    a. Long wheel life at high material removal rates
    b. Little or no thermal damage to the cutting edge because of the cool-cutting action
    c. Consistent sharp, burr-free cutting edges with no loss of hardness
    d. Easy and more effective control over tool sizes, shapes, and finishes
    e. Increased tool grinding productivity because of:
        (1) Less downtime due to wheel breakdown and conditioning
        (2) Less time required for gaging, sparkout, and wheel changes
    f. High operator acceptance because of the ease of grinding and the quality of the ground product
    g. High customer acceptance because tools ground with CBN wheels generally cut more cleanly and stay sharp much longer than those ground with conventional wheels.

Because of the exceptional hardness of CBN wheels, tool dimensions are accurately maintained with minimum downtime for wheel maintenance. Grinding with CBN wheels improves the fatigue strength and extends the useful life of the cutting tool.

EFFECTIVE GRINDING REQUIREMENTS

Before replacing an aluminum oxide wheel with a CBN grinding wheel it is important to make sure that the grinding system can take advantage of the productivity potential that a CBN wheel offers. Factors to be considered are the grinding machine, use of coolant, and horsepower requirements.

Fig. 7-38 CBN wheels work best on grinder with good spindle bearings and close-fitting slides. *(Courtesy of Cincinnati Milacron Co.)*

4. *Reliable feed rate* to preserve wheel life. This is *a must* for long wheel life and a good workpiece surface finish.

Coolant

Many operations on a tool and cutter grinder can be performed dry. By applying a constant supply of cutting fluid to the wheel–workpiece interface, however, the life of the CBN wheel can be extended by as much as ten times and produce a better surface finish.

Although straight oil performs best, a 10 percent solution of water-soluble oil generally works best for most operations. Use of water or water with a rust inhibitor is not recommended.

Horsepower Requirements

The materials ground with CBN wheels are usually hard, abrasion-resistant, and difficult to machine. These materials, compared to softer materials, require more horsepower to remove a given amount of material from the workpiece.

WHEEL SELECTION

Various types of CBN abrasives, in numerous grit sizes, are available so that the best CBN wheel can be selected for a given hardened-steel tool grinding application. For most tool and cutter grinding operations on hardened steel, a 100-concentration, 120-grit-size, resin-bond CBN wheel is recommended (Fig. 7-39). These wheels provide an excellent combination of high material-removal rates, good finishes,

Fig. 7-39 A 100-concentration, 120-grit-size, resin-bond CBN wheel is recommended for most tool-and-cutter grinding operations. *(Courtesy of GE Superabrasives)*

and long grinding-wheel life without generating excessive heat.

When selecting a CBN wheel for tool and cutter grinding operations, the following points must be considered:

Bond Type. The bond is a primary consideration in selecting the proper CBN grinding wheel. Resin-bond wheels are used for most tool and cutter grinder operations. Electroplated wheels are used when form grinding.

Grit Size. When replacing an aluminum oxide wheel with a CBN wheel it is wise to follow the recommendations listed in Table 7-6 regarding grit sizes.

Concentration. Wheels with 75 to 100 concentration are recommended because they provide good material-removal rates and long wheel life and are usually the most cost-effective.

NOTE Be sure to read Chapter 4, which gives a detailed explanation of the factors affecting CBN wheel selection. CBN wheel manufacturers' recommendations for their use should be closely followed.

WHEEL MOUNTING AND PREPARATION

Chapter 5 gives a detailed explanation of mounting and preparing a CBN wheel for the grinding operation. Be sure to refer to this chapter before proceeding with this operation. The following are intended as a reminder of the key points which must be covered if the grinding operation is to be successful:

1. The CBN wheel should be mounted on a high-quality adapter and should be kept together as a unit for the life of the wheel.
2. Use a dial indicator to true the wheel to within 0.001 in. (0.02 mm) or less runout on the wheel face.
3. Use the appropriate truing device for the type of CBN wheel and true the wheel on the machine on which it will be used (Fig. 7-40).
4. Dress the wheel to expose sharp CBN abrasive grains.

Table 7-6 COMPARATIVE GRIT SIZES

Aluminum Oxide Wheels	CBN Wheels
46–60	80–140
80–120	170–325
150–220	325–400

Fig. 7-40 If a Type 11V9 cup wheel is out of truth, a diamond-impregnated tool can be used to true the wheel. *(Courtesy of GE Superabrasives)*

NOTE Do *not* true or dress electroplated wheels.

GUIDELINES FOR GRINDING

Factors such as the grinding method, grinding mode, and the speed and feed rates used will affect the efficiency of a tool and cutter grinding operation.

Grinding Method

On tool and cutter grinders, the grinding method can be either wet or dry; both are used with success.

- *Dry grinding* with the resin-bond CBN wheel has proved to be effective (Fig. 7-41). Low feed rates are recommended to prevent workpiece burning.

- *Wet grinding* with either straight oil or a heavy-duty, water-soluble oil is very effective. It is important that the cutting fluid be applied directly to the work–wheel interface for the maximum cooling and lubrication.

Grinding Mode

If the grinder has sufficient power and rigidity, creep-feed single-pass grinding is recommended. Generally this mode of grinding produces the highest productivity, longest wheel life, and best workpiece finish. Conventional multipass grinding is also very effective for most tool and cutter grinder operations.

Wheel Speeds

In *dry grinding*, wheel speeds in the 3000 to 4500 sf/min (15 to 23 m/s) range are recommended. Higher wheel speeds may cause burning of the tool edges.

Fig. 7-41 Many operations can be performed dry on a tool-and-cutter grinder. *(Courtesy of GE Superabrasives)*

In *wet grinding*, wheel speeds in the 5000 to 6500 sf/min (25 to 33 m/s) range provide excellent results. The higher speeds improve both wheel life and metal-removal rates.

Feed

The traverse feed rate should be constant because of the cool, free-cutting characteristics of CBN grinding wheels. Whenever possible, use the creep-feed single-pass grinding mode. *Use lower feed rates when grinding dry.*

Roughing cuts should be about 0.002 (0.05 mm) deep, while finish cuts are usually 0.0005 to 0.001 in. (0.01 to 0.02 mm) deep. Sparkout passes are not necessary when using CBN wheels because if the grinder is in good condition, whatever is set for the depth of cut is what is removed from the tool being ground.

AUTOMATED END-MILL SHARPENING

The grinding of end mills requires skilled hands because there are multiple setups, multiple tool handlings, and a good chance for cumulative error. The result is that no two tools are ground exactly alike, no matter how high the operator's skill. Conventional grinding must be done dry so that the operator can hear and see what is happening. The use of coolant interferes with the sound of grinding and also makes it difficult to see what is happening. Also, in order to keep the cutter against the tooth rest, the grinding wheel is rotated into the end-mill cutting edge. This concentrates the heat at the cutting edge, and the teeth can lose some of their hardness due to thermal damage.

The Huffman grinder (Fig. 7-42) has been designed for the automated grinding of end mills.

- The program supplied with the machine makes it easy to specify the proper grinding specifications for various cutting tools.

- The programmer only enters basic information about the tool to be ground such as diameter, lead, number of flutes, length of flutes, primary and secondary clearance angles, and the size of the grinding wheel.

- The computer, using previously entered data, prints out the required program and the paper tape for use at the machine.

- The CNC on the grinder reads the tape once and loads it into the operating memory.

- Once the grinding cycle is started, all angles and clearances will be ground on the end mill in one setup.

ECONOMICS OF TOOL AND CUTTER GRINDING

The use of CBN grinding wheels for the grinding of end mills, milling cutters, hobs, and a variety of hardened cutting tools can reduce the cost of grinding and produce better-quality cutting edges. Since CBN wheels last much longer than do conventional grinding wheels and require little or no conditioning, there is less downtime for wheel maintenance. This results in increased productivity and reduced grinding costs. The savings in grinding costs with CBN wheels range from 20 to 50 percent or even higher on some applications. The hob shown in Fig. 7-43, previously sharpened in 2 h with an aluminum oxide wheel, took only 20 min with a CBN wheel.

Refer to Chapter 6 for a detailed explanation of the economics of using CBN grinding wheels.

Fig. 7-43 A CBN wheel resharpened this hob in 20 min, while the aluminum oxide wheel took 2 h. *(Courtesy of GE Superabrasives)*

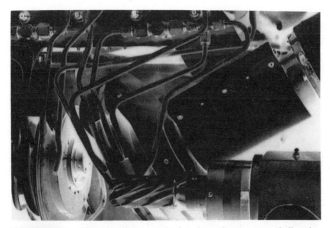

Fig. 7-42 The Huffman grinder has been specially designed to automatically grind milling cutters with CBN wheels. *(Courtesy of GE Superabrasives)*

1. What occurs when aluminum oxide abrasive dulls during tool-grinding?

CBN versus Aluminum Oxide Abrasive

2. List four advantages of using CBN wheels for grinding cutting tools.

3. Why are CBN wheels more productive than aluminum oxide wheels?

Effective Grinding Requirements

4. What characteristics must a tool and cutter grinder have to use CBN wheels effectively?

Wheel Selection

5. What wheel is recommended for most tool and cutter grinding operations? Explain why.

6. If a 46- to 60-grit aluminum oxide wheel is being replaced, what grit size should the CBN wheel have?

Wheel Mounting and Preparation

7. List the five important steps which must be considered for CBN wheels.

Guidelines for Grinding

8. What four factors can affect the efficiency of a grinding operation?

9. What wheel speeds are recommended for:
   a. dry grinding?
   b. wet grinding?

10. Why are sparkout passes not required when grinding with CBN wheels?

Automated End-Mill Sharpening

11. What features does the Huffman grinder have to grind cutters automatically?

Jig Grinding

The need for accurate hole locations in hardened work led to the development of the jig grinder. Often the clamping, the machining, or the hardening operation would distort the workpiece and alter the hole locations so that they were no longer accurate. Although the jig grinder was designed primarily for accurately locating holes in hardened workpieces, it has found wide use for the grinding of contour forms such as radii, tangents, angles, and flats (Fig. 7-44).

Conventional abrasive wheels and pins, although providing reasonable results, have a tendency to break down quickly, making it difficult to maintain hole size without constant measuring and compensating for wear. CBN abrasive wheels, which are twice as hard as aluminum oxide, resist breakdown and generally last much longer, requiring fewer adjustments and wheel changes.

OBJECTIVES

After completing this section you should be able to:

1. Select the proper CBN wheel or pin for the material to be jig ground

2. Mount, true, and dress the wheel or pin to obtain the best grinding results

3. Determine the proper speeds, feeds, and grinding mode for each job

CBN WHEELS FOR JIG GRINDING

Cubic boron nitride abrasive is almost twice as hard (4700 to 2000 KHN) and far more abrasion-resistant than aluminum oxide abrasive. Therefore, wheels and pins made with CBN generally outperform aluminum oxide wheels by wide margins when jig grinding difficult-to-grind (DTG) materials.

Cubic boron nitride wheels are designed to per-

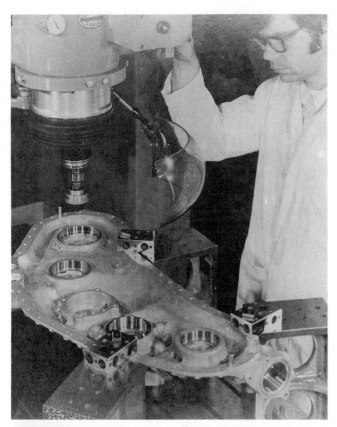

Fig. 7-44 Jig grinders are used for grinding holes and various contour forms to accurate size and location. *(Courtesy of GE Superabrasives)*

5. *Less Spindle Deflection.* CBN wheels stay sharp and free-cutting and remove stock through an efficient chip-making action. Therefore, there is less grinding pressure even at high material-removal rates and less spindle or wheel-shank deflection. This feature results in better part or hole geometry.

6. *Cooler Grinding.* Because CBN wheels are free-cutting, they also cut cool. There is little risk of burning the workpiece and causing subsurface thermal damage that could lead to the cracking of a tool, die, or critical part in service. Parts ground with CBN wheels have little or no metallurgical damage and therefore have a longer life in production.

7. *Faster Stock Removal.* The free-cutting, cool-grinding action of CBN wheels, along with their resistance to wear, make it possible to remove stock from 30 to 50 percent faster than with aluminum oxide wheels. This increases the productivity of the jig grinder.

8. *Reduced Grinding Costs.* Although CBN wheels cost more than do aluminum oxide wheels, they pay for themselves quickly by longer wheel life, shorter grinding time, and improved part quality. CBN wheels are cost-effective and show a much higher profit ratio than aluminum oxide wheels for jig grinding.

MACHINE SPECIFICATIONS

The performance of any grinding wheel depends on the capacity and working condition of the jig grinding machine. This is especially true with CBN wheels, and to take advantage of the productivity that these wheels offer, the jig grinder must have the following characteristics.

Rigid Spindle Bearings. Generally material is ground faster with CBN wheels than with aluminum oxide wheels, and with high material-removal rates the grinding forces are greater. Therefore, it is important that the spindle bearings be in good working condition. Worn bearings produce poor surface finishes and quickly reduce the life of the CBN wheel.

Spindle Speeds and Feeds. For the optimum grinding efficiency with CBN wheels, the grinding-wheel spindle should have a speed range of 3000 to 6000 sf/min (15 to 30 m/s).

The jig grinder should have fine outfeed control capability. A 0.0001- to 0.0002-in. (0.002- to 0.005-mm) increment outfeed is necessary to achieve the best performance.

The planetary offset function of the jig grinder should have a radial offset feature with a capability of reaching high speeds for small-hole jig grinding.

For chop grinding operations with CBN wheels, it

form best when grinding tool and die, carbon, and alloy steels hardened to Rc 50 and above, hard abrasive cast irons, and superalloys with hardness of Rc 35 and above.

Some of the advantages that CBN wheels offer over aluminum oxide wheels in jig grinding are:

1. *Longer Wheel Life.* CBN abrasive resists breakdown better than does aluminum oxide abrasive. As a result, CBN wheels last much longer than aluminum oxide wheels when grinding DTG materials. This results in fewer wheel changes with less interruption to production.

2. *Less Wheel Maintenance.* Unlike aluminum oxide wheels, CBN wheels rarely need to be retrued or reconditioned while running a job. Therefore, there are fewer interruptions in the grinding cycle.

3. *Positive Size Control.* Because CBN wheel breakdown is minimal, part size can be controlled with fewer in-process measurements and fewer stops to make machine adjustments to compensate for wheel wear. CBN wheels provide consistent size control over a long production run.

4. *Consistent Surface Finishes.* The CBN wheel's ability to resist breakdown produces predictable surface finishes from the start to the completion of a job.

is very important that the jig grinder have a rapid vertical reciprocation capability. A frequency of 200 to 250 strokes per minute is generally sufficient for most grinding applications.

> **NOTE** See Chapter 5 for a detailed explanation of the qualities a grinder must have to increase productivity, reduce machine downtime, and produce high-quality parts and consistent surface finishes.

WHEEL SELECTION

CBN jig grinding wheels are available in various bond types and a wide variety of styles to suit various jig grinding operations (Fig. 7-45). It is important that the proper wheel be selected to suit the workpiece material so that the most efficient grinding can occur. The most important factors to consider when selecting CBN wheels for jig grinding operations are the abrasive type, bond, and grit size.

Abrasive Type

A variety of CBN abrasives are available; however, the Borazon CBN Type I, Type II, Type 500 and Type 510 are the most common used for jig grinding operations. The application guide in Table 7-7 lists a few CBN abrasives and their applications. The CBN abrasives commonly used for jig grinding will be covered in the following subsection on bonds.

Bond

Cubic boron nitride jig grinding wheels are available in electroplated, resin-bond, vitrified-bond, and metal-bond types.

ELECTROPLATED WHEELS. These wheels have a single layer of CBN abrasive plated to a steel core or shank. Two exceptionally hard, tough types of Borazon CBN abrasive—Type I and Type 500—are recommended for electroplated wheels. Both of these abrasives are designed to provide low cost, high stock-removal rates, free cutting, and cool grinding for excellent performance.

RESIN-BOND WHEELS. Resin-bond wheels generally are made of Borazon Type II crystals which are uniformily distributed throughout the wheel matrix. These crystals are metal-coated for superior retention in the bond and last much longer than do electroplated wheels. Resin-bond wheels have high material-removal rates and long wheel life and produce consistent superior surface finishes.

VITRIFIED-BOND WHEELS. Vitrified-bond wheels, generally made from Borazon Type I abrasive, give essentially the same advantages as do resin-bond wheels. These wheels can be used immediately after truing and do not require conditioning.

METAL-BOND WHEELS. Borazon Type I abrasive is sometimes used for metal-bond wheels; however, Type 510 abrasive is specifically designed for use in metal bonds. Metal-bond wheels have long wheel life and excellent form retention. Metal-bond wheels with carbide shanks are recommended where wheel deflection is a problem. For form grinding, the largest wheel diameter possible should be used.

Grit Size

For a given set of operating conditions, a wheel containing coarse (large-grit-size) CBN abrasive will have long wheel life and faster material-removal rates than will a wheel with fine abrasive. The wheel with the finer grit size will generally produce higher surface finishes.

A good rule to follow when selecting abrasive grit size is to select the largest size that will produce the desired surface finish.

> **NOTE** See the detailed explanation in Chapter 4 regarding wheel selection, and be sure to follow the manufacturer's recommendations.

WHEEL MOUNTING AND PREPARATION

The importance of correctly mounting and preparing a CBN wheel for jig grinding cannot be overstressed. Correct mounting, truing, and conditioning are essential for the good performance of CBN jig grinding wheels.

Mounting the Wheel

The proper mounting of resin, vitrified-, or metal-bond wheels reduces the amount of abrasive which must be removed from the wheel to make it run true. Poor alignment shortens wheel life and affects work quality. The wheel overhang should be kept to a minimum to reduce deflection, chatter, and vibration.

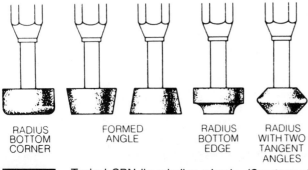

RADIUS BOTTOM CORNER   FORMED ANGLE   RADIUS BOTTOM EDGE   RADIUS WITH TWO TANGENT ANGLES

Fig. 7-45 Typical CBN jig grinding wheels. *(Courtesy of GE Superabrasives)*

Table 7-7

BORAZON CBN ABRASIVE APPLICATION GUIDE

Abrasive	Type I	Type 500
Wheel diameter	<⅛ in. (3 mm)	>⅛ in. (3 mm)
Conditions	Recommended for small pins where rigidity is inadequate for effective use of CBN Type 500	Recommended for systems where sufficient rigidity is maintained; CBN Type 500 is tougher than CBN Type I; higher feed rates are required to increase metal-removal rate

To ensure that a wheel is mounted in truth, check the wheel shank with a dial indicator (Fig. 7-46). The wheel should be slowly rotated by hand, and there should be no more than 0.001 in. (0.02 mm) runout. This is especially critical with electroplated wheels because they only have a single layer of abrasive crystal and excessive runout would shorten their life and reduce grinding efficiency.

Truing the Wheel

Only resin-, vitrified-, and metal-bond wheels require truing; *electroplated wheels are never trued or conditioned.* The recommended truing tool for resin-, vitrified-, and metal-bond wheels is a diamond-impregnated nib (Fig. 7-47). The diamond grit in the nib should be 150-grit. *Single-point diamond nibs should not be used because they may damage the CBN wheel.* Sharp or lapped single-point diamonds may be used to true a form on resin or vitrified-bond wheels; however, the diamond

must be kept *sharp* and small dress increments of 0.0005 in. (0.01 mm) or less should be used.

Always true CBN jig grinding wheels while applying a spray-mist coolant of water-soluble oil. This is critical when truing resin-, vitrified-, or metal-bond wheels, which overheat without coolant. Overheating can cause premature failure of the CBN wheel.

Cubic boron nitride wheels should be trued at fast feed rates [30 to 40 in./min (0.75 to 1 m/min)] (Fig. 7-48). Infeed increments should not be more than 0.0002 in. (0.005 mm) per pass.

Conditioning

Resin- and metal-bond wheels become glazed when trued and will not cut in this condition. To expose the sharp edges of the CBN crystals, it is necessary to condition (dress) the wheel. An aluminum oxide dressing stick (220 grit, medium hardness) saturated with water-soluble oil must be forced directly into the

Fig. 7-46 An indicator on the wheel shank being used to check the concentricity (runout) of the CBN jig grinding wheel. *(Courtesy of GE Superabrasives)*

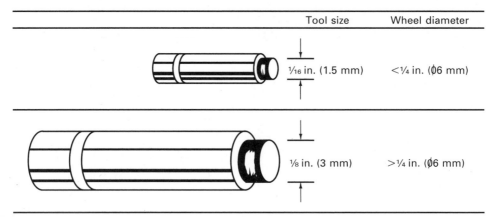

	Tool size	Wheel diameter
	¹⁄₁₆ in. (1.5 mm)	<¼ in. (ø6 mm)
	⅛ in. (3 mm)	>¼ in. (ø6 mm)

Fig. 7-47 Diamond-impregnated nibs are recommended for truing resin-, vitrified-, and metal-bond CBN wheels. *(Courtesy of GE Superabrasives)*

revolving wheel (Fig. 7-49). *Never use an up-and-down motion when conditioning a CBN wheel.*

As the dressing stick is forced into the glazed wheel, the stick will be consumed slowly. As the bond material is removed, sharp crystals are exposed and the stick is consumed rapidly. As soon as this occurs, the wheel is in good condition for grinding and further conditioning will simply waste the wheel material.

NOTE Vitrified-bond wheels seldom require conditioning, and electroplated wheels should not be trued or conditioned.

JIG GRINDING GUIDELINES

The efficiency of a jig grinding operation and the surface finish produced will also be affected by vari-

ables such as the wheel speed, reciprocal and planetary speed, grinding method, and the grinding mode.

Wheel Speeds

CBN wheels remove stock most efficiently and produce best surface finishes when they are operated at high speeds (3000 to 6000 sf/min or 15 to 30 m/s). Speeds below this range will shorten the life of the wheel and reduce the productivity of the grinding operation.

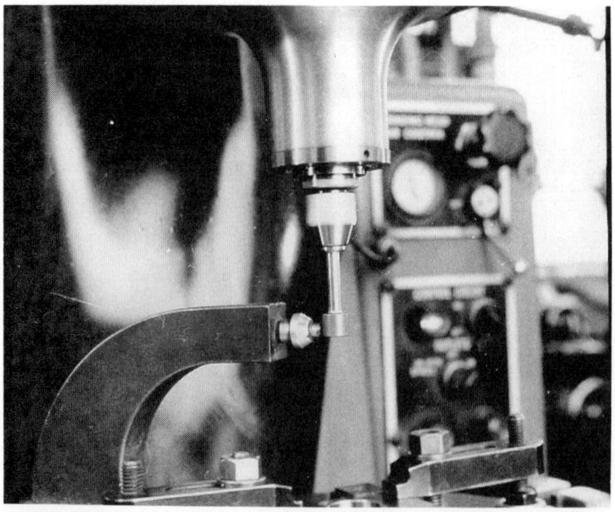

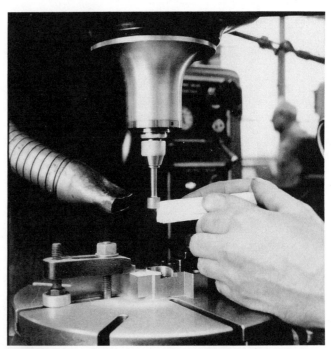

Fig. 7-48 When truing a CBN wheel, the wheel should be fed past a rigidly supported diamond-impregnated nib. *(Courtesy of GE Superabrasives)*

Fig. 7-49 Metal- and resin-bond CBN wheels should be conditioned with an aluminum oxide dressing stick after truing. *(Courtesy of GE Superabrasives)*

Never run CBN wheels faster than speeds that are recommended by the manufacturer.

Reciprocal and Planetary Speeds

For the best CBN wheel performance, use fast reciprocal and planetary speeds at continuous light outfeed rates (less than 0.0005 in. or 0.01 mm). The best outfeed rate for each job depends on the spindle speed, wheel diameter and bond, and workpiece material. Table 7-8 lists some recommended feed rates for electroplated CBN wheels.

Grinding Methods

The free-cutting and cool-grinding characteristics of CBN wheels usually make it possible to grind dry without excessive heating and thermal damage to the workpiece. Should heating become a problem, the wheel should be checked to determine whether it requires conditioning, the feed should be reduced, or a spray-mist solution of water-soluble oil should be directed to the wheel–work interface.

Grinding Modes

Several grinding modes are used for various jig grinding operations. These include hole or outfeed grinding, chop grinding, wipe grinding, and shoulder or bottom grinding.

The type of grinding mode will have an influence on the surface finish produced. For example, in wipe grinding the individual crystals may cut grooves in the workpiece along the lines of work travel. This does not happen in grinding modes where the wheel reciprocates as well as rotates.

Hole or Outfeed Grinding (Fig. 7-50) removes stock by a continuous outfeed of the wheel while grinding. The surface finish produced falls between wipe and chop grinding and is very sensitive to changes in the reciprocating rate and the wheel speed. The operator must use trial and error to arrive at the best material removal rates while producing the required surface finish. For maximum stock removal, use a coarse grit wheel; change to a finer grit to produce the desired surface finish.

Chop Grinding (Fig. 7-51) removes material by a re-

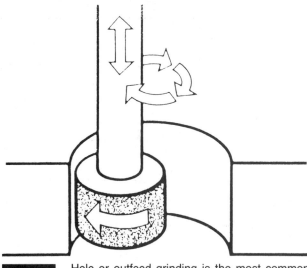

Fig. 7-50 Hole or outfeed grinding is the most common method of grinding holes. *(Courtesy of GE Superabrasives)*

ciprocating movement of a revolving wheel while the work is fed past it. This mode is similar to the action of a vertical shaper or slotter. A good surface finish can be produced by this grinding mode when using a coarse-grit wheel. For example, a 100 grit CBN abrasive wheel can produce a 20-RMS (0.05-μm Ra) surface finish. The chop grinding mode is excellent for contour grinding with a numerically controlled jig grinder.

Wipe Grinding (Fig. 7-52) removes stock by moving the workpiece past the revolving wheel; the wheel position is stationary. This grinding mode is used for flat and contour form grinding and produces the least favorable surface finish because any high-abrasive wheel particles will produce lines or grooves on the work surface. An electroplated wheel provides the best stock removal but produces a rough finish. A small grit-size resin-, vitrified-, or metal-bond wheel will produce good surface finishes.

Shoulder or Bottom Grinding (Fig. 7-53) removes material by a downfeed movement of the wheel into the workpiece. To achieve the best surface finish and size control, downfeed increments should be 0.0002 to 0.0003 in. (0.005 to 0.008 mm). A positive stop or a precision depth stop should be used to control the

Table 7-8	**CBN ELECTROPLATED WHEELS FEED RATES**
Wheel Diameter	**Outfeed Rate per Pass**
<0.060 in. (1.5 mm)	0.0001–0.0002 in. (0.003–0.005 mm)
0.060–0.156 in. (1.5–4.0 mm)	0.0003–0.0005 in. (0.008–0.012 mm)
0.156–0.375 in. (4.0–10.0 mm)	0.0006–0.001 in. (0.015–0.025 mm)
>0.375 in. (10.0 mm)	0.001–0.0015 in. (0.025–0.04 mm)

Note: Outfeed amounts per side. For finer finish, use a smaller grit size or reduced infeeds. These are only general guidelines. Some trial and error will be required to find the best operating parameters for a given grinding application.

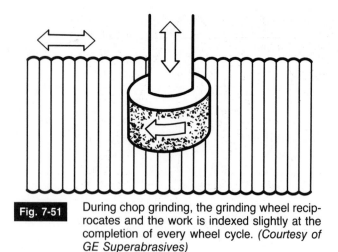

depth of the shoulder. A spray-mist coolant of water-soluble oil should always be used when shoulder or bottom grinding.

NUMERICAL-CONTROL JIG GRINDING

Continuous-path numerical-control (NC) jig grinding techniques require grinding abrasives that last a long time, retain their shape, and produce precise finishes and geometry without thermal damage to the workpiece. One factor not compensated for in the programmed jig grinder is wheel wear. If the wheel loses shape, size, or stock-removal capability while making a pass, an inaccurate contour will be produced.

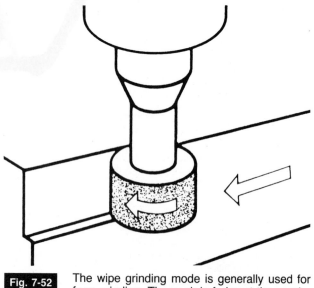

Fig. 7-52 The wipe grinding mode is generally used for form grinding. The work is fed past the revolving wheel, which is in a stationary position. *(Courtesy of GE Superabrasives)*

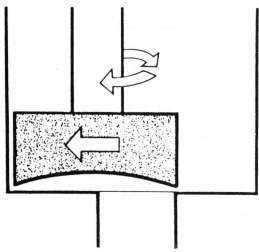

Fig. 7-53 The bottom of the wheel must be relieved when shoulder or bottom grinding. *(Courtesy of GE Superabrasives)*

CBN wheels provide the precise high stock removal required in economic NC jig grinding (Fig. 7-54). With light outfeeds and fast spindle, reciprocating and planetary speeds, CBN makes most difficult jig grinding easy to perform. The varied grit sizes and types of CBN abrasive gives the operator a wide selection to fit even the most complex contoured job (Table 7-9).

NC Jig Grinding Advantages

Because of its hardness and wear resistance, CBN abrasive is finding wide use in NC jig grinding applications. The grinder can be programmed with predictable results which cannot be achieved with aluminum oxide wheels. Some of the benefits of using CBN wheels for jig grinding are:

- Dies and punches are mated with proper clearance.

- Wheel wear is minimal; there is no difficulty in blending tangent points.

- Close tolerances can be maintained when grinding large areas such as cams.

- Parts that require superior finishes, such as contour gages, can be ground to finishes as fine as 10 RMS. These finishes are usually obtained by chop grinding.

- With predictable results, the operator learning time is shortened.

- Deep small holes can be ground to precise specifications.

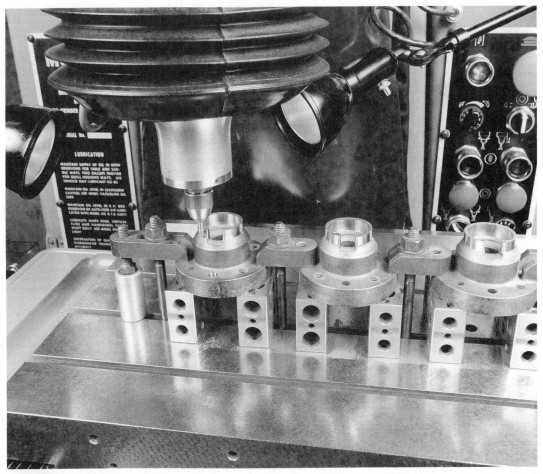

Fig. 7-54 Numerical-control jig grinding is an ideal application for CBN wheels. *(Courtesy of GE Superabrasives)*

Table 7-9 CBN JIG GRINDING CHECKLIST

Unsatisfactory Condition	Possible Cause	Corrective Action
1. Poor finishes	Abrasive grit size too coarse	—Use finer grit size —Increase wheel speed —Decrease infeed —Decrease feed rate
	Chatter	—Check wheel runout —True vitrified wheel —True and condition (resin and metal bond wheels only)
2. Bellmouth	Wheel overtravel Wheel deflection	—Shorten stroke length —Reduce wheel overhang
3. Tapered parts	Vertical slide or quill misalignment	—Realign slide or quill
4. Poor cutting Part burning	Wheel loading	—Condition wheel (resin and metal bond wheels only)
5. Short wheel life	Low wheel speed Excessive runout	—Increase wheel speed —Check wheel runout —True and condition (resin-, metal-, and vitrified-bond wheels only)

1. For what purpose are jig grinders used in the metalworking industry?
2. List four reasons why CBN wheels are used for jig grinding.

CBN Wheels for Jig Grinding

3. For what type of steels are CBN wheels suited?
4. List the eight advantages CBN wheels have over aluminum oxide wheels in jig grinding.
5. Why do parts ground with CBN wheels have little or no metallurgical damage?

Machine Specifications

6. Name two requirements which a jig grinder must have to take advantage of the productivity of CBN wheels.

Wheel Selection

7. What CBN abrasives are commonly used for jig grinding wheels or pins?
8. Name the four bond types used in CBN jig grinding wheels.
9. What bond types have the following characteristics or purposes?
   a. one layer of abrasive
   b. produce consistent surface finishes
   c. long life and excellent form retention
10. What is the rule to follow when selecting abrasive grit size?

Wheel Mounting and Preparation

11. State the causes of:
    a. poor alignment
    b. long wheel overhang
12. How close should a CBN wheel be trued with an indicator?
13. What type of tool should be used for truing resin-, vitrified-, and metal-bond CBN wheels?
14. Why should a spray-mist coolant be used when truing CBN wheels?
15. What type of dressing stick should be used to condition resin- and metal-bond wheels?

Jig Grinding Guidelines

16. Name the variables which affect the efficiency of a jig grinding operation.
17. What would be the effect of running CBN wheels below the recommended range?
18. Why can CBN wheels grind dry without excessive heating and damage to the workpiece?
19. What is the most common method of jig grinding holes?
20. For what purpose are the following grinding modes used:
    a. chop grinding?
    b. wipe grinding?

Numerical-Control Jig Grinding

21. List the possible causes for the following unsatisfactory conditions:
    a. poor finishes
    b. short wheel life
    c. part burning

Honing

For more than half a century, the abrasives most commonly used for honing were aluminum oxide and silicon carbide. These conventional abrasives performed well but could remove only small amounts of stock and were not satisfactory on very hard materials. Superabrasive honing stones, made from CBN or diamond, are achieving material-removal rates many times that possible with conventional honing stones. Almost any type of material can be successfully honed by one or both of these superabrasives. This includes all steels, alloys, superalloys, cast iron, carbides, platings and coatings, ceramics, and glass.

After completing this section you should be able to:

1. Select the proper hone or stone for the material to be honed
2. Set the proper speeds and feeds for honing operations
3. List the advantages of honing with CBN hones

DEVELOPMENTS IN HONING

Honing is a low-speed surface-finishing operation, usually on a bore or internal diameter, in which stock is removed by the shearing action of abrasive grains. The honing action consists of a simultaneous rotary and reciprocal motion of a hone body which carries a number of abrasive hones or "sticks" (Fig. 7-55). This action produces a characteristic criss-cross pattern on a bore, known as *crosshatch*, which helps to retain oil and provides an excellent bearing surface for mating parts. This process not only improves the surface finish and integrity of holes but also ensures that they are straight, round, and to the correct size.

While honing and honing machines have been in use since early in this century, the process has remained relatively unchanged for most of this time. A major development in honing was the advent of superabrasives. In 1957 synthetic diamond and in 1969 cubic boron nitride (CBN) became commercially available. Considerable research by the General Electric Company (GE) followed, and in the early 1970s Borazon CBN Type 510, an extremely hard and tough abrasive, was developed for the metal-bond grinding, and honing, of hard tool steels and superalloys.

The development of CBN abrasive by GE and specifically Borazon CBN Type 510 crystal has revolutionized the honing process. Tests conducted in the laboratory and in industry showed that honing stones made with CBN abrasive easily outperformed hones made with conventional abrasives such as alu-

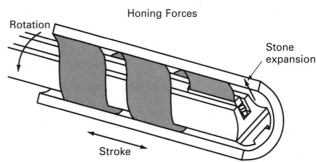

Fig. 7-55 A honing operation produces a crosshatch pattern on an internal diameter through the reciprocal and rotational action of the hone. *(Courtesy of Sunnen Products Co.)*

minum oxide. They hone faster, last up to 100 times as long, and hone without the high-pitched squeal common to the honing operation. Honing with CBN abrasives produces consistent and predictable results while improving workpiece geometry (Fig. 7-56).

HONING VERSUS GRINDING

There are basic differences between honing and grinding which must be recognized (Fig. 7-57). Honing is a low-speed operation (85 to 300 sf/min or 25 to 95 m/min), while grinding is generally a high-speed operation (5000 to 6500 sf/min or 25 to 33 m/s).

• Chips produced by *grinding* are short, hot sparks due to the intermittent contact of each abrasive particle on the workpiece surface.

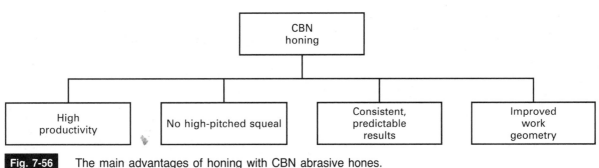

Fig. 7-56 The main advantages of honing with CBN abrasive hones.

Grinding Versus Honing

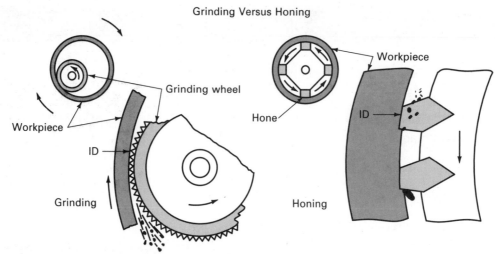

Fig. 7-57 Honing is a low-speed method, while grinding is a high-speed method of finishing internal diameters (ID). *(Courtesy of GE Superabrasives)*

- A continuous cooler chip is produced by honing due to the continuous contact of the abrasive hone with the work (Fig. 7-58).

- Grinding with aluminum oxide tends to damage the surface of the workpiece, often to a depth of up to 0.002 in. (0.05 mm).

- Honing is a gentle, cooler finishing operation, and thus there is little chance of damaging or distorting the workpiece surface.

- Honing requires no chucking of the workpiece because small parts float and align themselves automatically on the expanding hone.

- Honing with CBN hones is much more cost-effective than grinding on most internal operations.

- Honing produces holes having a good surface finish and high accuracy in size, parallelism, and roundness faster than by grinding.

Fig. 7-58 The correct application of CBN stones produce a continuous chip. (Chips shown actual size.) *(Courtesy of Wickman Corporation)*

COMMON BORE ERRORS

Honing is required wherever bore errors interfere with the fitting or the operation of a part. Bore errors (Fig. 7-59), commonly caused by heat-treating, chucking, and machining, include undersize, out-of-roundness, bellmouthing, waviness, boring tool and chatter marks, barrel-shaped and tapered holes, rainbow shapes, and misalignment. The honing operation is capable of correcting these common bore problems faster and more accurately than grinding, without changing the location of the bore or damaging the work surface.

THE HONING PROCESS

An abrasive stone, mounted on a mandrel (Fig. 7-60), is reciprocated (moved back and forth) in the bore as the mandrel rotates. During this honing action each CBN crystal cuts a chip from the workpiece. As the chip is continuously formed, it wears a crater in the bond in front of the CBN crystal. As the honing stone is reciprocated in the bore, the chip is carried around the sides of the abrasive crystal forming V-shaped chip trenches (Fig. 7-61). These trenches allow the chip to exit at the rear of the honing stone.

Cubic boron nitride honing stones generally have relatively narrow widths. If the stones are wide, the chips cannot escape and tend to pack between the hone–work interface, which occasionally causes deep scratches in the bore. With the controlled wearing away of the bond, good sharp crystals are exposed, which provides high honing ratios and long stone life.

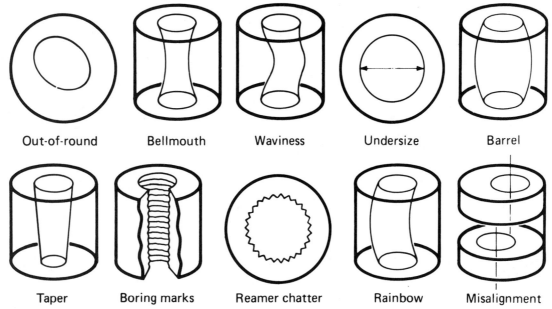

Out-of-round	Bellmouth	Waviness	Undersize	Barrel
Taper	Boring marks	Reamer chatter	Rainbow	Misalignment

Fig. 7-59 Honing can correct common bore errors created during the manufacture of metal parts. *(Courtesy of Sunnen Products Co.)*

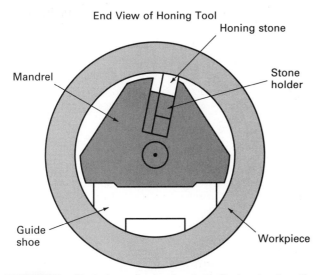

End View of Honing Tool

Honing stone

Mandrel

Stone holder

Guide shoe

Workpiece

Fig. 7-60 End view of a honing tool. During honing, the hone, mounted in a mandrel, is moved back and forth in the bore while it is rotating. *(Courtesy of Sunnen Products Co.)*

Fig. 7-61 The chip generated during honing wears a crater in the bond in front of the crystal. *(Courtesy of GE Superabrasives)*

Types of Honing Processes

There are two general types of honing processes in industrial use today; *conventional* and *single-stroke* honing.

CONVENTIONAL HONING. During conventional honing three forces are applied to the hone to produce the cutting action:

1. *Cutting pressure* (expansion) is applied when the mandrel forces the stone and the guide shoe surfaces into contact with the bore.
2. The *rotational and reciprocal action* of the hone in the bore causes thousands of small cutting edges on the stone to shear minute chips from the workpiece.
3. Each chip is a long strand of metal because the hone is constantly in motion and in contact with the bore (Fig. 7-58).

SINGLE-STROKE HONING. The single-stroke honing tool (Fig. 7-62) consists of an expandable diamond-plated abrasive sleeve on a tapered arbor which is expanded to size by a calibrated adjusting screw.

- It is faster and maintains more consistent size accuracy than conventional honing.

- There is only one stroke (pass through the workpiece), and the size is set only once.

- Single-stroke honing is limited to finish honing operations.

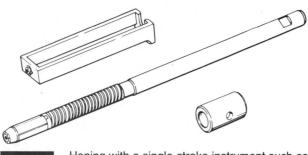

Fig. 7-62 Honing with a single-stroke instrument such as this offers both speed and accuracy over conventional honing. *(Courtesy of Sunnen Products Co.)*

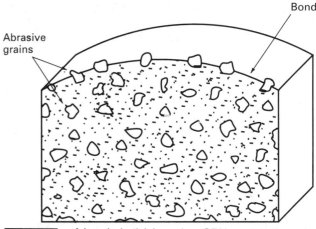

Fig. 7-63 A bonded-stick hone has CBN crystals throughout the bond matrix. *(Courtesy of Sunnen Products Co.)*

TYPES OF HONES

There are two categories of abrasive hones; *conventional* and *superabrasive*.

1. *Conventional abrasive hones*, made of aluminum oxide or silicon carbide, are used primarily for honing soft steel, cast iron, and nonferrous metals.
2. *Superabrasive hones*, made of diamond or CBN abrasive, are used for honing hard and soft steels.

   a. *Diamond hones* are used mainly for honing carbides, ceramics, and glass. They are not suitable for honing steel because diamond reacts chemically with steel and will break down quickly.
   b. *CBN hones* are widely used for honing steels, hardened steels, and superalloys. They have proved to be more efficient and cost-effective in the honing of mild steel. Since CBN hones are twice as hard as aluminum oxide hones, they last up to 100 times as long in mild steel with a much lower noise factor.

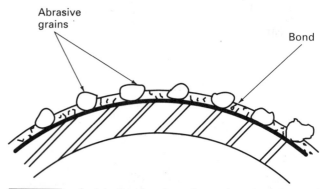

Fig. 7-64 A plated surface hone has only a single layer of CBN crystals attached to the surface of the hone by an electrolytic plating process. *(Courtesy of Sunnen Products Co.)*

Types of Superabrasive Honing Stones

There are two basic types of honing stones or "sticks"—*bonded sticks* and *plated surfaces*.

1. A *bonded stick* (Fig. 7-63) consists of a bond matrix which holds the abrasive grains together. As the outer layer of abrasive grains wear, they break away from the bond and expose new sharp grains.
2. Hones with *plated surfaces* (Fig. 7-64) consist of a single layer of CBN abrasive grain attached to the surface by an electrolytic plating process. The bond in plated hones is generally chromium, cobalt, nickel, or copper.

Bonded abrasive sticks are made with three types of bonds—metal, resin, and vitrified.

Metal bonds may be a mixture of metal powders which are generally iron- or copper-based or bronze.

- Metal bonds are very wear-resistant and can be designed to wear at the same rate as the abrasive of the hone. This process keeps the hone sharp and free-cutting.

- Because of the strong retention bond between CBN crystals and the metal matrix, these hones are used when high metal-removal rates are desired.

- Due to the strength of the bond, high honing speeds (250 to 300 ft/min or 75 to 95 m/min) are possible and desirable with these stones.

- Metal-bond stones are extremely durable and are often used for finishing rough bores.

Resin bonds are composed of organic materials such as resin or plastic. Resin-bond stones are made with Borazon CBN Type II (coated monocrystalline) abrasive and are generally used for low metal-removal rates and noninterrupted cuts.

- They produce excellent finishes depending on the grit size and the concentration.
- Resin-bond stones are freer cutting than metal bond stones but not as wear-resistant or as long-lasting.

Vitrified bonds are similar to the bonds used in conventional abrasive grinding wheels. Since this bond is composed of fused clay, it is much more fragile than a metal bond and is used mainly in finishing operations.

- These stones perform well on hard steels and wear-resistant alloys.
- Because vitrified stones can contain a higher concentration of CBN than other bonds, they produce excellent finishes.

NOTE When selecting the proper bond hardness for any of the three bonds mentioned, remember the rule: *Use a hard bond on soft materials and a soft bond on hard materials.*

Honing Stones

There are various types of honing stones and mandrels available to suit a variety of honing applications.

Single-stone hones (Fig. 7-60) usually have three unequally spaced contact points, the stone and two guide shoe points, to maintain contact with the cylinder wall. This unequal spacing reduces the chance of chatter during the honing process. Single-stone hones are generally used on small bores [less than 3 in. (75 mm)].

Four-point-contact hones (Fig. 7-65), which consist of two hones and two guide shoes, are capable of faster metal removal but are not as accurate as the three-point-contact hone. Four-point-contact hones are generally used to finish large-diameter bores [over 3 in. (75 mm)].

Multi-point-contact hones (Fig. 7-66) may contain as many as eight abrasive sticks along with an appropriate number of guide shoes. These hones are generally used on production for the honing of large parts.

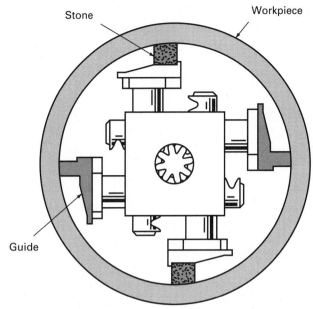

Stone Workpiece

Guide

Fig. 7-65 Four-point-contact hones are used to hone large diameters. *(Courtesy of Sunnen Products Co.)*

MACHINE REQUIREMENTS

Most existing manual-stroke and power-stroke honing machines work well with CBN hones provided they are in good condition. To obtain the fullest potential that superabrasives such as CBN offer, the machine should have the following qualities:

- Be rigid and with no vibration in the spindle
- Have sufficient power to permit high material-removal rates
- Have high spindle speeds—about three times the speed for conventional hones
- Have closely controlled increments of stone feed or pressure
- Have a coolant system capable of delivering adequate volumes of filtered honing fluids

The success of a honing operation depends on such factors as the grit size and concentration of the abrasive, cutting speed, feed rate, and honing pressure.

Grit Size

A honing stone is selected on the basis of several variables: grit size, hardness, bond type, and abrasive material.

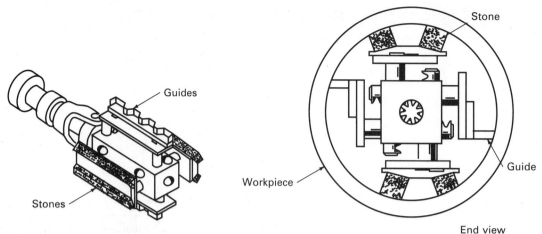

Fig. 7-66 Multipoint-contact hones are used on production for honing large parts. *(Courtesy of Sunnen Products Co.)*

- Grit size can be determined by looking up the surface finish chart in Table 7-10 and choosing the size that will produce the desired surface finish.

- Hardness can be selected by referring to the manufacturer's catalog. If a stone does not cut, use a softer grade; if it wears too fast, use a harder grade.

- Roughing cuts of 0.012 to 0.020 in./min (0.3 to 0.5 m/min) can be achieved with coarse-grit CBN abrasive having a low concentration.

- For finish cuts and close dimensional tolerances, stock allowance should be from 0.002 to 0.008 in. (0.05 to 0.20 mm). A fine-grit abrasive with high concentration should be used for finish honing.

Refer to Table 7-11 for recommended stone specifications and operating conditions for honing hard steels (>58 Rc).

NOTE For general-purpose honing, an 80- to 100-grit, metal-bond, 35-concentration stone will provide high material-removal rates and long hone life.

Cutting Speed

The cutting speed of honing with superabrasives is normally twice that normally used for aluminum oxide hones. Reciprocal or stroke speed must be proportional to the spindle speed to obtain the desired crosshatch pattern. *High spindle speed is critical for honing with CBN stones; without it, honing is less effective.*

Table 7-10 APPROXIMATE SURFACE FINISH IN MICROINCHES (μ-IN.) RA

Material	Abrasive type	Grit Size								
		80	100	150	220	280	320	400	500	600
Hard steel	Aluminum oxide, silicon carbide	25	—	20	18	12	10	5	3	1
	CBN	—	55	45	30	28	—	20	—	7
Soft steel	Aluminum oxide, silicon carbide	80	—	35,55*	25	20,35*	16	7,10*	4,8*	2
	CBN	—	65	—	—	—	—	25	—	—
Cast iron	Silicon carbide	100	—	30,40*	20	12	10	6	5	3
	Diamond	—	—	—	80	—	—	50	—	20
Aluminum, brass, bronze	Silicon carbide	170	—	80	55	33	27	15	12	2
Carbide	Diamond	—	—	30	20	—	—	7	—	2
Ceramic	Diamond	—	—	50	40	—	—	20	—	15
Glass	Diamond	—	—	95	70	—	—	30	—	15

*The first number is for small parts, honed on machines with 1 hp or less; the second number is for large parts, honed on machines with 2 hp or more.
Source: Sunnen Products Co.

Table 7-11	OPERATING CONDITIONS FOR HONING HARD STEELS
Rough Hone—High Material-Removal Rate	
Finish:	>40 RMS (8.0 μm Rt)
Grit:	100
Concentration:	50
Bond/Abrasive:	Metal-Borazon CBN 510
Speed:	200-300 sf/min
Stone width:	⅛-¼ in. (3–6 mm)
Finish Hone—Low Material-Removal Rate	
Finish:	<20 RMS (4.0 μm Rt)
Grit:	300 (or finer depending on finish required)
Concentration:	75–150
Bond/Abrasive:	Resin-Borazon CBN Type II
	Metal-Borazon CBN Type 510
	Vitrified-Borazon CBN Type I
Speed:	125–300 sf/min
Stone width:	⅛-¼ in. (3–6 mm)

Spindle Speed

- Use 115 to 300 sf/min (35 to 95 m/min) for high material-removal rates on mild and hardened steels. Use high honing pressures.

- For finish honing, use 66 to 197 sf/min (20 to 60 m/min) and reduce the honing pressure by about 50 percent.

Reciprocal or Stroke Speed

For most honing operations, a stroke speed of 200 to 300 in./min (8 to 12 m/min) is recommended. This may have to be varied somewhat to produce the desired crosshatch pattern in the bore.

Coolants

A good honing oil is very important to a successful honing operation. The honing oil serves three functions:

1. It cleans and lubricates the honing stones.
2. It carries away the heat and waste produced by the cutting action.
3. It prevents galling between the guide shoe and the internal work surface.

Straight oil, mineral seal, or soluble oil is preferred when honing with CBN. Some synthetic fluids work well, although water or water-soluble synthetics sometimes cause problems for CBN hones.

To be effective, a honing oil should have the following characteristics:

- Be inactive at normal temperatures so that it does not corrode.

- Become instantly active when the temperature comes close to the melting point of the metal being honed.

   –This high temperature occurs in microscopic spots at the point of the cutting action and causes welding of the metal-guide shoe to the metal being honed.

   –These tiny weld spots may be torn apart by the honing action, causing rough surface finishes and rapid wear of the hone and guide shoe.

- Prevent welding by chemically changing the hot spots from metallic to nonmetallic compounds which cannot weld. This problem is very common when stainless steels are being honed.

NOTE The use of the proper honing oil should not be neglected. Some oils work better than others on certain jobs. Always use the oils recommended by the hone manufacturer.

TRUING AND CONDITIONING

Honing stones have flat edges when purchased and will not function efficiently because only the corner of the stone would be in contact with the internal diameter. Before use, a stone must be trued by one of the following methods:

Truing

METHOD 1. They can be "broken in" on the machine and workpiece on which they will be used. The hone is operated at low speeds and low pressure until the working surface of the hone is worn into full contact with the bore of the workpiece. This can be a very time-consuming operation because of the wear resistance of CBN stones.

METHOD 2. To produce the required roundness and cylindrical form on the abrasive layer, the honing tool should be set up and ground to the proper radius on a cylindrical grinder (Fig. 7-67). A vitrified-bond, K-hardness, silicon carbide wheel with about the same grit size as the honing stones should be used for the truing operation. The grinding wheel should be operated at a slow speed (200 to 500 sf/min or 60 to 152 m/min) using a trickle of grinding fluid to build up an abrasive slurry.

Fig. 7-67   A CBN hone being trued to the required roundness and size with a silicon carbide grinding wheel. *(Courtesy of Gehring L. P. Co.)*

Conditioning

After truing and whenever the honing stone does not cut effectively, the abrasive crystals no longer project from the wheel bond. This glazed or loaded condition can be removed by blasting the working surface of the hone with a fine grit size of aluminum oxide or silicon carbide abrasive. Figure 7-68 shows a honing stone before and after conditioning.

HONING GUIDELINES

To select the proper CBN hone for the job, you must consider the

- Type of work material
- Hardness of the material
- Amount of material to be removed.

The following general rules apply to any CBN hone, regardless of the bond. (Refer to Fig. 7-69.)

1. For maximum removal rates, use coarse grit sizes and lower concentration.
2. For maximum material removal rates, use narrow stones (⅛ to ¼ in. or 3 to 6 mm) at high speed. Use only sufficient pressure to assure free cutting.
3. For fine finishes, use fine grit sizes and higher concentrations.

NOTE   The stone may be wider if finish is the first consideration and metal removal is secondary.

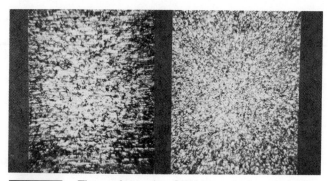

Fig. 7-68   The surface of a CBN hone before and after conditioning. *(Courtesy of Gehring L. P. Co.)*

4. Use the highest speed possible which will give long stone life and high metal-removal rates. CBN stones are generally more effective at higher honing speeds.
5. Use the lowest machine pressure which will permit free cutting.
6. Interrupted cuts require higher concentrations of abrasives and/or harder bonds.
7. Always use an adequate supply of coolant. High material-removal rates require sufficient coolant to cleanse the stone, clear the chip, and cool the workpiece.
8. For hard materials, use a soft bond; for soft materials, use a hard bond.

For more specific information regarding stone selection for honing hardened and mild steels, see Table 7-12. Some of the most common honing problems and their causes and possible remedies are covered in Table 7-13.

ADVANTAGES OF SUPERABRASIVE HONES

The advantages of CBN honing stones over the conventional abrasives are listed in Fig. 7-70.

- *Higher Material Removal.* Material-removal rates are up to twice those of conventional abrasives.

- *Slow Wear.* Because aluminum oxide stones wear rapidly, the honing ratio (the volume of the material removed to the volume of the stone consumed) is in the range of 5 to 10. The honing ratio for CBN stones is in the range of 800 to 1200.

- *Better Bore Accuracy.* CBN stones remain free cutting at low machine pressures. Aluminum oxide stones, on the other hand, are less free-cutting and require higher operating pressures, which may result in bellmouthing or barreling of the bore.

- *Reduced Heat.* CBN stones produce less heat

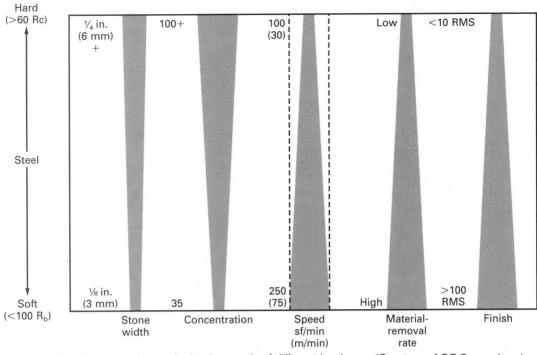

Fig. 7-69 General guidelines for honing steels of different hardness. *(Courtesy of GE Superabrasives)*

than do aluminum oxide stones because of the decrease in cutting pressure.

• *Less Power Consumption.* With CBN hones all the power is used to remove material from the workpiece. With aluminum oxide, much of the power is used to achieve stone breakdown.

• *Uniform Honing.* Honing time is predictable because the structure and the cutting action of each CBN stone are consistent. Aluminum oxide stones often lack uniformity from stone to stone and must be changed to suit various honing conditions.

Table 7-12   CONDENSED GUIDELINES FOR HONING WITH CBN STONES

	Hardened Steels	Mild Steels
Stone type	Metal-bond	Metal-bond
Bond hardness Rockwell "B"	93–98	103–110
Borazon CBN abrasive type	510, 550	510, 550
Abrasive concentration	50–100	35
Abrasive grit size	Roughing: 100 Finishing: 300	100
Stone width, in. (mm)	0.13–0.25 in (3–6 mm)	0.13–0.15 in (3–4 mm)
Honing speed, sf/min (m/min)	High MRR:* 115–295 (35–90) Best finishes: 66–180 (20–60)	≥115–295 (35–90)
Crosshatch angle	40–70°	40–70°
Specific pressure, N/mm²	High MRR: 2–4 Best finishes: 1–2	(2–4)
Honing fluid	Straight oil	Straight oil

*Material-removal rate.

Table 7-13 **HONING TROUBLESHOOTING CHECKLIST**

Problem	Cause	Remedy
Stone glaze	Peripheral speed too low	Increase spindle speed
	Stone pressure too low	Increase pressure
	Abrasive concentration too high	Be certain abrasive concentration is less than 50, preferably 35
Low honing ratio	Excessive material-removal rates	Reduce pressure first; if wear is still excessive, reduce spindle speed
Long time required for stone break-in	Worn honing head	Use a well-conditioned head
Excessive heat	Excessive material-removal rates	Reduce stone pressure and/or spindle speed
	High friction (stone glazed)	Dress stone open, then follow procedures for "stone glaze" above
	Inadequate coolant flow	Increase coolant flow

- *Noise Reduction.* The loud squeal often found with aluminum oxide stones at high metal-removal rates is not present with CBN stones.

- *Improved Performance.* Performance is improved for honing hardened steel. A test performed on hardened pump cylinders showed a reduction of honing time by 50 percent and an increase in bore accuracy from 0.002 to 0.0005 in. (0.05 to 0.012 mm).

- *No Hone Reversal.* No hone reversal is required to clear or "open up" the stone as with alumi-num oxide stones when they become glazed. CBN metal-bonded stones are free-cutting and do not require reversing.

ECONOMIC EVALUATION

Because of the advantages of using CBN, it is now apparent that mild steel tubing can be honed faster and at a lower cost than previously possible with conventional abrasives. Figure 7-71 illustrates a typical honing cost analysis, showing the economic advantage of CBN hones over conventional hones.

In addition to the savings gained by using CBN, benefits such as improved workpiece geometry, efficient cutting action, excellent surface finishes, longer tool life, and noise reduction are realized. Although CBN honing stones cost more initially, the increase in productivity and the reduction in the scrap rate makes CBN hones cost-effective. Generally, honing with CBN results in lower total honing cost per piece.

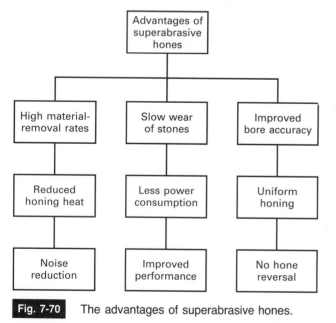

Fig. 7-70   The advantages of superabrasive hones.

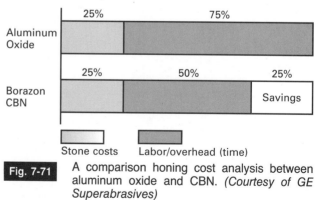

Fig. 7-71   A comparison honing cost analysis between aluminum oxide and CBN. *(Courtesy of GE Superabrasives)*

Developments in Honing

1. Describe the honing action.
2. What pattern is formed on the bore surface, and what purpose does it serve?
3. What development has revolutionized the operation of honing?
4. List five reasons why CBN abrasive crystals are far superior to conventional abrasives in the honing operation.

Honing versus Grinding

5. Compare the speeds used for honing and grinding.
6. Compare the chips produced by honing and grinding and state how these may affect the surface of a workpiece.

Common Bore Errors

7. List eight common bore errors which can be corrected by honing.

The Honing Process

8. Describe the action of a chip produced by CBN abrasive and state its purpose.
9. What forces are applied during conventional honing?
10. Describe a single-stroke honing tool.
11. For what purpose are the following hones used:
    a. diamond?
    b. CBN?
12. Describe the bonded stick and plated surface hones.
13. Name the three types of bonds used in hones and state where each is used.

Machine Requirements

14. List five requirements that a machine should have to use CBN hones effectively.
15. What hone is recommended for general-purpose honing?
16. What type of hone should be used for:
    a. roughing cuts?
    b. finishing cuts?
17. What are the recommended honing speeds for:
    a. spindle speed?
    b. reciprocal or stroke speed?
18. State three functions of a good honing oil.

Truing and Conditioning

19. Briefly describe the best method of truing a honing stone.
20. How can glazed or loaded honing stones be conditioned?

Honing Guidelines

21. What general rules apply for the following:
    a. maximum removal rates?
    b. fine surface finishes?
    c. long stone life?
    d. bond selection?

Advantages of Superabrasive Hones

22. Explain the advantage of CBN hones over conventional abrasives regarding the following:
    a. material removal rate
    b. honing ratio
    c. reduced heat
    d. noise reduction
    e. hone reversal

CHAPTER EIGHT

Cemented Carbide Tool Grinding with Diamond Wheels

One of the largest uses of diamond abrasives in the metalworking industry is in the manufacture and regrinding of cemented carbide tools and cutting-tool materials in toolrooms throughout the world. Diamond wheels are far more effective than silicon carbide wheels in the grinding of cemented carbide tools. They grind carbide tools with such speed and precision that they have become the accepted standard for grinding carbide tools throughout the world.

Diamond wheels, used extensively on tool and cutter grinders, may also find use on surface, cylindrical, internal, and jig grinders and honing operations. Since the widest use of diamond wheels is found on tool and cutter grinders, this chapter will deal mainly with this machine tool.

OBJECTIVES

After completing this chapter you should be able to:

1. Select the proper diamond wheel for grinding cemented carbide tools
2. Prepare the grinder and the diamond wheel for grinding
3. Compare the advantages and cost-effectiveness of diamond grinding wheels

DIAMOND ABRASIVES

The hardness of diamond abrasive is compared to aluminum oxide abrasive, silicon carbide abrasive, and cemented tungsten carbide tool material in Fig. 8-1. Since tungsten carbide is almost as hard as aluminum oxide abrasive, it is impossible to use it for grinding tungsten carbide. Silicon carbide, which is slightly harder than tungsten carbide, has been used in the past for grinding carbide tools. Some silicon carbide grinding wheels may still be used for this purpose, although diamond grinding wheels are much more cost-effective when the grinding ratios of these two wheels are compared (Fig. 8-2).

If the diamond and silicon carbide grinding ratio values were used in the grinding equation in Chapter 6, it would be quite clear that regrinding cemented carbide with silicon carbide abrasive is very expensive. Because diamond has over three times the hardness and ten times the abrasion resistance of silicon carbide, it has much higher material removal capabili-

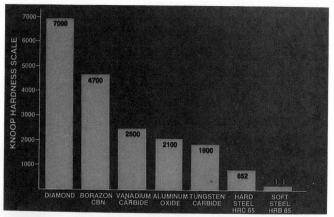

Fig. 8-1 A hardness comparison of abrasives and materials. *(Courtesy of GE Superabrasives)*

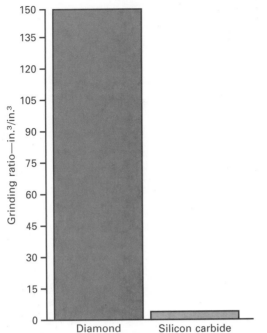

Fig. 8-2 Grinding ratios of silicon carbide and diamond wheels for grinding cemented carbide tools. *(Courtesy of GE Superabrasives)*

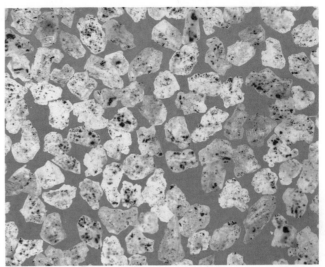

Fig. 8-3 Regular friable manufactured diamond is used primarily for regrinding tungsten carbide tools. *(Courtesy of GE Superabrasives)*

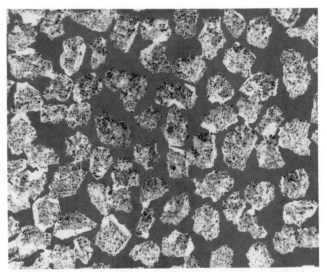

Fig. 8-4 Nickel-coated friable diamonds are used in resin-bond wheels to hold the crystals better and provide longer wheel life. *(Courtesy of GE Superabrasives)*

ties as compared to that of silicon carbide abrasive.

Since the characteristics of diamond abrasive can be controlled by the manufacturing process (see Chapter 2), diamond abrasive can be specially made to suit certain grinding applications or materials. The most widely used types of diamond (listed in Table 8-1) are described as follows:

1. The *friable manufactured diamond* (Type RVG) (Fig. 8-3) has a rough surface as well as an irregular shape. This diamond type is very free-cutting and is used primarily in toolrooms for regrinding (wet or dry) tungsten carbide cutting tools.
2. The *nickel-coated friable manufactured diamond* (Type RVG-W) (Fig. 8-4) is essentially the same as the regular friable diamond with an added nickel coating. The coating provides an improved holding surface for a resin bond which results in longer wheel life. This is the most widely accepted diamond type for grinding cemented carbides.

Table 8-1	DIAMOND ABRASIVES FOR GRINDING CEMENTED CARBIDES			
Abrasive	**Bond**	**Grit Sizes**	**Abrasive Description**	**Use**
RVG	Resin	40–500	Irregularly shaped friable crystals	Grinding tungsten carbide and ceramics under light feed conditions
RVG-W	Resin	40–500	Nickel-coated friable crystals	Grinding tungsten carbide under medium to severe conditions
RVG-D	Resin	40–500	Copper-coated friable crystals	Dry-grinding tungsten carbide
CSG-II	Resin	60–400	Nickel-coated, blocky crystals of medium toughness	Grinding cemented tungsten carbide–steel combinations

3. The *copper-coated friable manufactured diamond* (Type RVG-D) (Fig. 8-5) is basically the same as the regular friable diamond with a 50 wt % copper coating. This coating improves the mechanical strength of the diamond crystal and helps to control fracturing under the stresses caused by the high-material-removal-rate grinding of cemented carbides. This diamond abrasive has increased the grinding ratio threefold over regular uncoated diamond in dry grinding tungsten carbide. Since this crystal has harder characteristics, it should be used on grinders with higher power to the spindle.

4. *Special, blocky, nickel-coated diamond* (Type CSG II) (Fig. 8-6), used in a resin-bond system, is especially suited for grinding brazed carbide tools. In regrinding such tools, it is often necessary to grind some of the steel shank which supports the tungsten carbide insert. This tougher diamond resists wear in this particular type of operation.

Diamond grinding wheels have been accepted as the standard for grinding cemented carbide tools because they provide:

• Long wheel life at high material-removal rates

• No thermal damage to the cutting edge of the tool because of their cool-cutting action

• Consistent burr-free cutting edges with no loss of hardness

• Effective and precise control over tool sizes, shapes, and surface finishes

• Increased productivity because there is less downtime due to wheel breakdown and conditioning, gaging, sparkout passes, and wheel changes

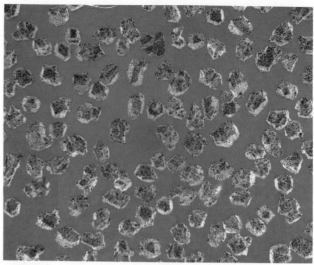

Fig. 8-6 Special, blocky, nickel-coated diamond can be used for wet or dry grinding some high-speed steels. *(Courtesy of GE Superabrasives)*

• High operator acceptance because of the ease of grinding and the quality of the product produced

• Wide customer acceptance because the tools ground stay sharp longer and produce clean, accurate cuts

CEMENTED CARBIDE TOOLS

Diamond wheels are widely used in the metalworking industry for grinding and reconditioning cemented carbide tools. Cemented carbides are designed for high hardness, toughness, and resistance to abrasive wear. These characteristics improve the performance of metal-cutting and metal-forming tools, but they make cemented carbides extremely difficult to grind.

Manufacture of Cemented Carbides

All cemented carbides are manufactured products. They are made by mixing micron-sized carbide and cobalt powders, compacting the mixture in a mold and sintering the resulting compact at a temperature high enough to cause the cobalt to flow. During sintering, the cobalt fills the void between the grains, thoroughly coating each grain. When the cobalt solidifies, it "cements" or binds the grains, forming a dense composite.

Abrasion Resistance

Cemented carbides get their hardness from the carbide grains and their toughness from the tight bonds produced by the cementing action of the cobalt metal. This combination of high hardness and toughness creates high abrasion resistance.

Fig. 8-5 Copper-coated diamonds are used for dry-grinding tungsten carbide tools. *(Courtesy of GE Superabrasives)*

There are many grades (compositions) of cemented carbides, each designed for a specific range of applications. Grades designed for high abrasion resistance—needed for cutting cast irons and for dies—contain varying percentages of tungsten carbide (WC) and cobalt (Co). Grades designed for cutting steels also contain titanium carbide (TiC), tantalum carbide (TaC), or both, giving them higher hot hardness and resistance to cratering wear.

Carbide grains are visible in the electron-beam micrograph of a cemented carbide used for steel-cutting tools (Fig. 8-7). Grindability is influenced by the relative percentage of tungsten carbide, titanium carbide, tantalum carbide, and cobalt and also by the size of the carbide grains, the thermal conductivity of the material, and its modulus of resilience—a measure of the ability of a material to absorb energy before failure.

The influence that the grade of carbide has on the life of a diamond wheel is shown in Table 8-2.

Diamond Wheels for Carbide Tools

Grinding a cemented carbide is a much more severe operation than grinding a ductile steel. When the steel is ground, the predominant material-removal mechanism is failure by shearing. Material is removed in the form of ductile chips, with relatively low use of energy and a relatively low rate of wheel wear.

When cemented carbides are ground, the material-removal mechanism is entirely different. The abrasive

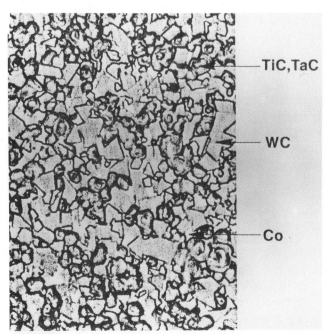

Fig. 8-7 The grindability of carbide is influenced by the amount of tungsten, titanium, or tantalum carbide which it contains and the size of the carbide grains. *(Courtesy of GE Superabrasives)*

in the grinding wheel fractures the tungsten carbide grains on the surface of the workpiece. The grinding swarf consists of a fine powder rather than chips.

Impact with the carbide destroys conventional grinding abrasives quickly. The Knoop indentation hardness of tungsten carbide grains is approximately 1880 kilograms per square millimeter (kg/mm^2), while the Knoop hardness of silicon carbide grains is only around 2480 kg/mm^2. These grains wear very fast when pitted against carbide grains, which are nearly as hard.

Diamond abrasive grains have a Knoop hardness of 7000 to 10,000, so they are about three to four times harder than silicon carbide grains. For this reason, diamond wheels are almost universally used for grinding cemented carbides.

EFFECTIVE GRINDING REQUIREMENTS

Before replacing a silicon carbide wheel with a diamond grinding wheel, it is important to make sure that the grinding system can take advantage of the productivity potential that a diamond wheel offers. Factors which must be considered are the grinding machine, use of coolant, and horsepower requirements.

Grinding Machine

The performance of a diamond wheel will vary with the condition of the grinder. A superabrasive grinding wheel will not work properly, if at all, on a grinder in poor condition. For best results when grinding with diamond wheels, the grinder must have the following characteristics:

1. *Tight Spindle Bearings.* This eliminates vibration and chatter and lengthens the life of the grinding wheel, resulting in excellent surface finish and dimensional accuracy of the cutting tool.
2. *Close-Fitting Slides.* These prevent chatter, which would reduce wheel effectiveness.
3. *Steady Spindle Speed.* Variations in spindle speed reduce cutting efficiency and shorten wheel life. Steady spindle speeds of 3000 to 4000 r/min, along with high torque capacity, are necessary for high metal-removal rates.
4. *Reliable Feed Rate.* This is a *must* for long wheel life and a good workpiece surface finish.

Horsepower Requirements

Cemented tungsten carbide tool materials are among the hardest and toughest materials manufactured. They are as tough as tool steels but are much harder. A comparison of the power required to grind carbide with diamond, and CBN to grind high-speed steel, is found in Fig. 8-8. As can be seen, the diamond wheel requires more than three times the energy to grind

Table 8-2

INFLUENCE OF CARBIDE GRADE ON WHEEL LIFE-WET SURFACE GRINDING-TYPE 1A1 WHEELS

Carbide Grade	Type of Application	Composition, %				Relative Grinding Ratios
		WC	Co	TiC	TaC	
883	Metal removal; cast irons, aluminum alloys	94	6	—	—	340
779	Metal forming	91	9	—	—	225
44A	Metal removal; cast irons, aluminum alloys	94	6	—	—	270
55A	Components	87	13	—	—	205
370	Metal removal; steel	72	8.5	8	11.5	145
350	Metal removal; steel	64	6	25.5	4.5	260

carbide as it requires for a CBN wheel to grind high-speed steel.

WHEEL SELECTION

Basically, a grinding wheel is a toolholder. The tools are crystals (grains) of abrasive, embedded in a bond or matrix on the wheel rim, which is supported by a core that fits the grinder spindle. When grinding cemented carbides, the physical properties of the diamond and the bond that holds the abrasive crystals in the wheel rim are very important. Using the wrong diamond abrasive or the wrong bond can result in reduced wheel life, low metal-removal rates, and increased grinding costs (Table 8-3).

Friable forms of diamond abrasive are recommended for wheels used to grind cemented carbides. A friable diamond crystal will microfracture under the stresses set up by dull cutting points, creating fresh, sharp points. These self-sharpening diamond abrasives, in the right bond, can grind cemented carbide tools efficiently, with good wheel life.

Resin bonds are recommended for wheels used to grind cemented carbides. Resin bonds are resilient, cushioning the diamond abrasive against shock, which might cause a large fracture. The resin bond is designed to wear or erode at the same rate as the diamond wears; therefore, sharp new diamond crystals are exposed when needed. As a result, resin-bonded diamond grinding wheels remain free-cutting and remove material up to ten times faster than metal-bonded wheels and twice as fast as vitreous-bonded wheels.

Wheel Selection Guidelines

Any successful grinding operation depends to a large extent on choosing the right wheel for the job. The type of wheel selected and how it is used will affect the material-removal rate (MRR) and the life of the grinding wheel. The selection of a diamond grinding wheel can be a complex task, and it is always wise to follow the manufacturer's suggestions for each type of wheel. They have had successful experience in designing and applying wheels for specific jobs, and therefore their suggestions usually result in selecting the best wheel for each job.

When selecting the proper diamond wheel for each job, it is important to consider factors such as concentration, wheel diameter, diamond depth (rim), and grit size, since all of these affect the material-removal rate and the grinding wheel life.

1. *Diamond Concentration.* High diamond concentrations (see Chapter 4) affect wheel life when grinding cemented carbides wet or dry. Figure 8-9, on the basis of tests using wheels with 150-grit diamonds, shows that the grinding ratio increases as the diamond concentration increases.
2. *Wheel Diameter.* Large-diameter wheels outlast small diameter wheels by a wide margin when cemented carbides are ground under the same conditions (Fig. 8-10). There is more diamond

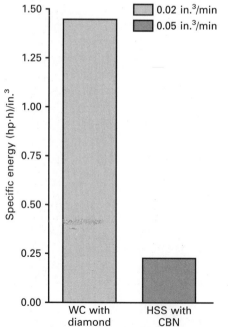

Fig. 8-8 A comparison of power required to grind cemented carbide with diamond and CBN wheels. *(Courtesy of GE Superabrasives)*

Table 8-3 **RESIN DIAMOND ABRASIVE SELECTION GUIDELINES**

Type of Abrasive	Dry Grinding		Wet Grinding	
	Cemented Carbide	Cemented Carbide/Steel	Cemented Carbide	Cemented Carbide/Steel
RVG	Fair	Good	Good	Good
RVG-W 56	Good	Better 0–15% steel	Better	Better
RVG-W 30	Good	Best overall	Good	Good
RVG-880	Better	No	Best	No
RVG-D	Best	No	No	No
CSG-II	No	Best over 33% steel	No	Best

abrasive to share the work in a large wheel; therefore, wheel life and overall cost effectiveness is improved.

3. *Diamond Depth.* The diamond depth (rim width) of a cup wheel used for tool and cutter grinding affects the wheel performance. Increasing the diamond depth from ¹⁄₁₆ in. (1.6 mm) to ⅛ in. (3.2 mm) almost doubles the grinding ratio (Fig. 8-11). If this change is combined with an increase in diamond concentration from 50 to 100, a sixfold increase in grinding ratio results in the best cost-effective grinding.

4. *Grit Size* (Fig. 8-12)
   a. Wheels with coarse diamond abrasive generally have longer life when grinding cemented carbides, under heavy feed conditions or on steel/carbide combinations, but require higher spindle power and generate more heat when grinding. Because of this, coarse diamond abrasive (80 to 120-grit) is generally more suited for wet grinding applications.

b. Wheels with fine diamond abrasive (150 and finer) are usually freer cutting, require less spindle power, and generate less heat when grinding. Wheels with fine diamond abrasive are generally recommended for dry grinding applications.

c. If the proper grit size for any material is not known, wheels containing 150 grit generally perform well under wet or dry grinding conditions, giving good surface finish on the ground part with reasonably good wheel life.

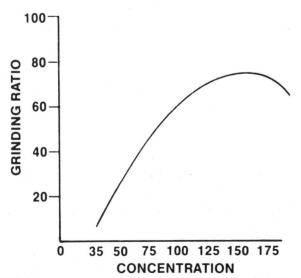

Fig. 8-9 The diamond concentration of a wheel affects the grinding ratio. (*Courtesy of GE Superabrasives*)

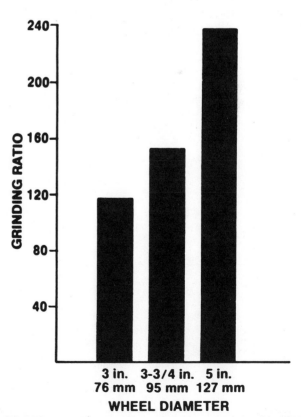

Fig. 8-10 Large-diameter wheels outperform small-diameter wheels by a wide margin. (*Courtesy of GE Superabrasives*)

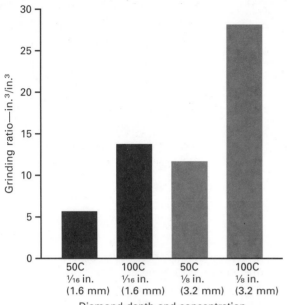

Fig. 8-11 A sixfold improvement in grinding ratio can result if diamond depth and concentration are combined. *(Courtesy of GE Superabrasives)*

WHEEL MOUNTING AND PREPARATION

Care should be taken whenever a diamond wheel is mounted because it pays off in longer wheel life and consistent accuracy in the parts ground. For a diamond wheel to grind at 100 percent efficiency, the wheel must run as true as possible.

Runout of a diamond wheel will cause it to chip on the edges, wear faster, produce poor surface finishes, and produce inaccurate parts. Wheels where the outside diameter is used for grinding should never run out more than 0.0005 in. (0.01 mm). Cup grinding

wheels should never run out more than 0.0005 in. (0.01 mm) on the grinding face. Careful attention in mounting the wheel will reduce initial wheel loss due to truing and dressing—or even eliminate this loss. Chapter 5, on preparing the CBN wheel and grinder, deals extensively with the mounting, truing, and dressing of CBN wheels. Many of the same points apply to diamond wheels; therefore, it is wise to refer to Chapter 5 before mounting, truing, and dressing diamond wheels.

The following is a summary of the key points which must be followed when mounting a diamond wheel if it is to perform satisfactorily:

1. The diamond wheel should be mounted on a high-quality adapter and should be kept together as a unit for the life of the wheel.
2. Mount the wheel-adapter unit on the grinder spindle and lightly tighten the flange nuts.
3. Check the runout on the grinding portion of the wheel with a dial indicator and correct the runout by:
   a. *Wheel Circumference.* Move the wheel on the adapter by gently tapping with a hammer on a wooden or plastic block until the runout is no more than 0.0005 in. (0.01 mm) (Fig. 8-13).
   b. *Wheel Face.* Face wheels are usually trued by the manufacturer and should have very little runout. If a runout greater than 0.0005 in (0.01 mm) is indicated, the wheel must be trued (Fig. 8-14).
4. Tighten the flange nuts securely and recheck the wheel runout.

Wheel Preparation

The condition of the grinding surface of a diamond wheel will determine the success or failure of any

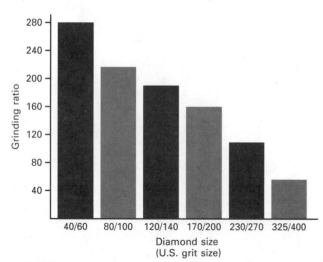

Fig. 8-12 Using the coarsest grit size possible results in the lowest grinding wheel cost per piece. *(Courtesy of GE Superabrasives)*

Fig. 8-13 Correcting the runout on the circumference of a diamond grinding wheel. *(Courtesy of Cincinnati Milacron Co.)*

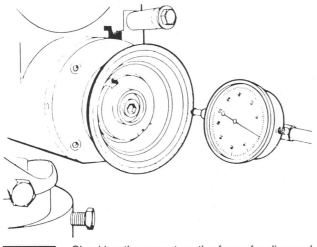

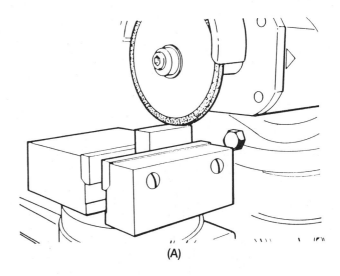

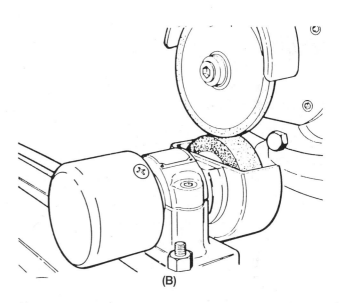

Fig. 8-14 Checking the runout on the face of a diamond cup wheel. *(Courtesy of Cincinnati Milacron Co.)*

grinding operation. Wheel conditioning, next to speeds and feeds, is the most important factor in the efficient use of diamond wheels. This can involve the operations of truing and dressing. For best performance, grinding wheels should be trued anytime they are removed from the grinder.

Truing

If the wheel runout is more than 0.0005 in. after mounting, it will be necessary to true the wheel. Truing consists of grinding or wearing away a portion of the abrasive section of the grinding wheel to make it run true or to bring it to the desired shape. Chapter 5 covers the operations of truing and dressing of CBN wheels in detail and should be followed for diamond wheels. The major difference between the two wheels will be that the diamond wheel is much harder and therefore will take longer to true and dress than is necessary for a CBN wheel.

Some of the more common methods of truing a diamond wheel are:

1. *Mild Steel Block* (Fig. 8-15A)
   a. Mount a mild steel block on the grinder.
   b. Take a few passes across the steel block at 0.001-in. (0.02-mm) depth per pass.
   c. Stop the grinder and check the wheel face or circumference for truth with a dial indicator.
   d. If necessary, continue passes over the steel block until the wheel runout is eliminated.
2. *Brake-Type Dresser* (Fig. 8-15B)
   a. Mount a brake-type dresser on the grinder.
   b. Follow the dresser manufacturer's recommendations on the use of this equipment.
   c. Continue the operation until the wheel runout is corrected.

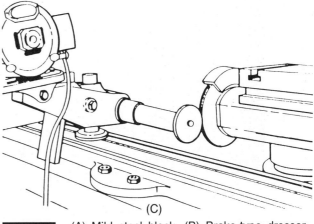

Fig. 8-15 (A) Mild steel block. (B) Brake-type dresser. (C) Toolpost grinder. *(Courtesy of Cincinnati Milacron Co.)*

3. *Toolpost Grinder* (Fig. 8-15C)

 a. Mount a toolpost grinder on the grinder table.

 b. Mount an 80- to 120-grit vitrified silicon carbide wheel on the grinder spindle to operate at about one-quarter of its normal speed.

 c. Set the diamond wheel and the silicon carbide wheel to operate in opposite directions.

 d. Take a few passes at 0.001 in. (0.02 mm) over the diamond wheel.

 e. Check the wheel for runout with a dial indicator.

 f. Continue taking passes over the diamond wheel until the runout is corrected.

Dressing

The truing process generally leaves the grinding surface of a diamond wheel smooth, with few or no abrasive crystals protruding above the wheel surface. A wheel in this condition would burn the workpiece and remove little or no work material. Dressing removes some of the bond material from the surface of a trued wheel to expose the diamond crystals and make the wheel grind efficiently.

To dress a diamond wheel:

1. Select a silicon carbide or aluminum oxide dressing stick.

2. Saturate the dressing stick with coolant so that a slurry is created when it contacts the diamond wheel.

3. Hold the dressing stick firmly and bring it into contact with the revolving diamond wheel (Fig. 8-16).

4. Feed the dressing stick aggressively into the wheel to open up the wheel face.

5. Once the dressing stick starts to wear rapidly, the diamond wheel is dressed.

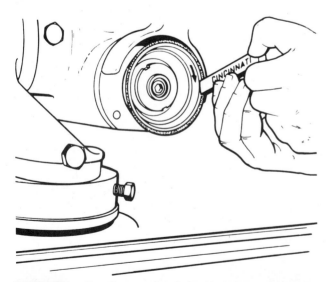

Fig. 8-16 Opening up the face of a diamond grinding wheel with a silicon carbide dressing stick. *(Courtesy of Cincinnati Milacron Co.)*

USING DIAMOND WHEELS

The proper use of diamond grinding wheels will result in long wheel life and high material-removal rates. To achieve the best grinding performance, the diamond wheel must be used under conditions which make this possible. Operating conditions such as wheel speed, work speed, feed rate, and the use of coolant are factors which affect the performance of a diamond grinding wheel.

Wheel Speed

In wet surface grinding, the speed of a diamond wheel is important to the grinding performance. Low wheel speeds (below 4000 sf/min) tend to reduce grinding efficiency, while speeds (over 6000 sf/min) tend to reduce wheel life. The best average wheel speed for most applications is in the 4000 to 6000 sf/min (20 to 30 m/s).

The best wheel speed for dry tool and cutter grinding is 3500 sf/min (18 m/s), using medium to fine grit sizes (150 and finer) at 75 or 100 concentration. Many tool and cutter grinders are set to drive a wheel at 3500 to 4500 sf/min (18 to 23 m/s). *Do not dry grind at wheel speeds over 4500 sf/min (23 m/s).* Dry grinding at speeds over 4500 sf/min (23 m/s) will cause heat damage to the diamond wheel. If it is absolutely necessary to operate at above 4500 sf/min (23 m/s), a coolant should be used to prevent damage to the wheel and extend its life.

Work Speed

Table traverse speed or crossfeed is largely governed by the amount of downfeed or depth of cut. Because these can vary greatly depending on the machine condition, grinding operation, workpiece material, and other factors, it is difficult to give specific work speed rates. The best work speed, often arrived at through trial and error, is where there is no loss of wheel speed, excessive wheel wear, or damage to the wheel or work.

Feed Rate

Excessive feed or depth of cut will shorten the life of a diamond grinding wheel and should be avoided. Too deep a cut can result in burned and cracked wheels, chipped or cracked carbide, and failure to obtain part size, finish, and form.

In face grinding cemented carbides, trying to take a heavy cut too quickly will result in the carbide acting as a shear and dressing the diamond wheel face. When grinding on the periphery of the wheel, heavy cuts cause the wheel to climb the workpiece, causing the work to dress the wheel out of round and producing chatter.

The depth of cut possible with diamond wheels is governed by the diamond grit size. The recommended depths of cut for various diamond wheel grit sizes are listed in Table 8-4. Feeds more than those recommended will cause material of the part to get under and cut away many diamond grits per revolution that will never get a chance to cut.

Cutting Fluids

Cutting fluids should be used whenever possible when using diamond wheels to reduce the grinding heat and extend the wheel life. If it is necessary to grind dry, as in tool and cutter grinding, a resin-bond wheel should be used. Sometimes spray mist with a rust inhibitor or refrigerated air can be used as a coolant to reduce the grinding heat; however, this is not as effective as wet grinding.

A small trickle of coolant occasionally applied is worse than no coolant at all. This causes alternate heating and quenching, which can cause damage to the diamond wheel and carbide tools. The best way to apply coolant is to allow the centrifugal force of the wheel to deliver it in a steady stream to the point of grinding contact (Fig. 8-17).

Today, there is a wide range of resin-bonded grinding wheels which have been specifically designed to be used dry when grinding cemented carbide tools. These wheels are very effective and maintain productivity without the use of any coolant.

Carbide Grinding Economics

When grinding cemented carbides with diamond wheels, two sets of costs must be considered—labor and overhead cost per piece and diamond wheel cost per piece. Labor and overhead cost per piece can be reduced by increasing the material-removal rate. This increases the diamond wheel cost per piece since diamond wheel wear increases with the material-removal rate.

Generally, the economic material-removal rate—the rate at which labor and overhead cost per piece and wheel cost per piece are balanced for lowest total grinding cost—is quite low when grinding cemented carbides in comparison with steel grinding. For dry tool and cutter grinding, the optimum material-

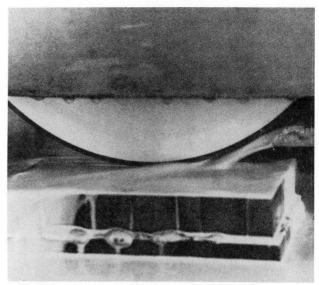

Fig. 8-17 A flood coolant directed to the grinding zone can increase wheel life ten times over dry grinding. *(Courtesy of GE Superabrasives)*

removal rate is seldom more than 1 to 2 in.^{3} (16.4 to 32.8 cm^3) of cemented carbide per hour. For wet surface, cylindrical, and vertical spindle grinding, the optimum material-removal rate usually ranges from 2 to 10 in.^{3} (32.8 to 164 cm^3) of cemented carbide per hour.

To realistically evaluate the economics of cemented carbide grinding operations, it is necessary to know how well the wheel is performing. A useful index of diamond wheel performance is the *grinding ratio*. The grinding ratio is obtained by dividing the volume of workpiece material removed in a given time by the volume of wheel consumed in the same period, when operating under constant conditions. The higher the grinding ratio, the longer the life of the wheel (Fig. 8-18).

Material-Removal Rates

With the right wheel for the job, running at the right speed, a variety of material-removal rates are possible. Low material-removal rates extend wheel life at

Table 8-4	RECOMMENDED DEPTH OF CUTS FOR DIAMOND GRIT SIZES			
Diamond Grit Size	Grit Diameter		Recommended Maximum Depth of Cut	
	in.	mm	in.	mm
100	0.006	0.15	0.002	0.05
150	0.004	0.10	0.0015	0.04
220	0.003	0.08	0.0008	0.02
400	0.0015	0.04	0.0004	0.01

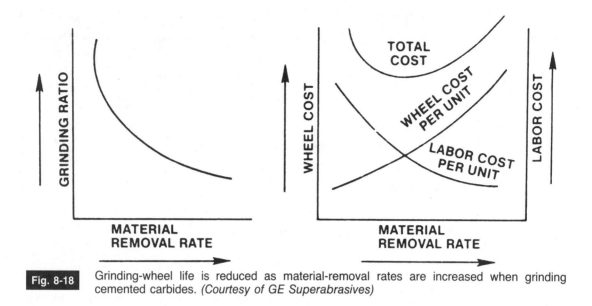

Grinding-wheel life is reduced as material-removal rates are increased when grinding cemented carbides. *(Courtesy of GE Superabrasives)*

the expense of productivity. High material-removal rates increase productivity at a sacrifice of some wheel life. In most industries, material-removal rates are selected on the basis of least total grinding cost per piece. Generally, this is at a point where wheel cost per piece and labor-and-overhead cost per piece are approximately equal.

There is no clearly defined method of increasing material-removal rate that will guarantee the most effective use of the diamond grinding wheel. Laboratory tests have indicated that in both wet and dry grinding cemented carbides, the life of the wheel is shortened in some direct relationship to the increase in material-removal rate, irrespective of the technique employed for increasing removal rate. However, there is some indication that in the case of wet surface grinding the use of heavy crossfeeds will take a somewhat lesser toll in wheel life than will downfeed or table speed.

However, a very important factor to consider is that the wheel is removing carbide only while the wheel and the workpiece are in contact. Any technique which increases the noncontact time between the wheel and the workpiece diminishes the effective material-removal rate and increases cost.

Wheel Life

The life of a diamond grinding wheel can be affected by the wheel speed, the work speed, the depth of cut, the grinding mode, the area of wheel contact, and the type of carbide ground. All these factors have an effect on the forces which are created during the grinding process.

Three forces act on a diamond wheel in tool and cutter grinding of cemented carbides: normal force, tangential force, and radial force (Fig. 8-19). *Normal force* is by far the greatest of the three forces which act

on a wheel. Increasing the infeed (depth of cut) from 0.002 to 0.004 in. (0.05 to 0.10 mm) reduces the wheel life because the increased pressure will cause the wheel to break down more quickly. The three forces are interrelated, and the specific grinding conditions will determine the amount of each force component.

TOOL AND CUTTER GRINDING GUIDELINES

The tool and cutter grinder is the most common machine tool used for regrinding or reconditioning carbide and high-speed steel-cutting tools. This machine is versatile and provides the operator with a good view of the sharpening operation. For best results in regrinding cutting tools, the following guidelines are offered:

1. *Reduce the Area of Contact.* The contact surface on the wheel face should not be much more than ⅛ in. (3 mm) wide. Narrow wheel rims are suitable for most grinding jobs. Although wider wheel rims give longer wheel life, they create more grinding heat.
2. *Use Low Wheel Speeds.* Wheel speeds should be around 3500 sf/min (18 m/s) for dry grinding operations in order to reduce the heat created at the cutting tool edges.
3. *Use Reduced Work Speed.* The table speed should not be more than 6 to 9 ft/min (2 to 3 m/min) to reduce the possibility of diamond grit being torn out of the wheel due to crowding. Diamond wheels will perform better with deeper infeeds and slower work (table) traverse speeds.
4. *Depth of Cut.* The depth of cut is governed by the diamond grit size. For a 100-grit diamond wheel, the maximum depth of cut should be 0.002 in. (0.05 mm) and 0.0004 in. (0.01 mm) for a 400-grit wheel.

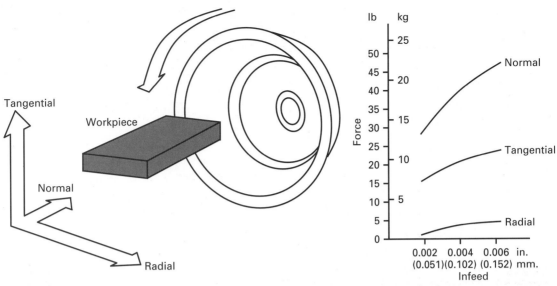

The forces which act on a diamond wheel in tool-and-cutter grinding of cemented carbides. *(Courtesy of GE Superabrasives)*

REVIEW QUESTIONS

1. Where is the largest use of diamond abrasives in the metalworking industry?

Diamond Abrasives

2. What is the Knoop hardness scale reading of tungsten carbide, silicon carbide, and diamond?
3. Describe and state the use of the following diamond abrasives.
   a. regular friable
   b. copper-coated friable
   c. blocky, nickel-coated
4. List five important reasons why diamond grinding wheels have been accepted as the standard for grinding cemented carbide tools.

Cemented Carbide Tools

5. For what purpose are cemented carbides designed?
6. List four things that influence the grindability of cemented carbides.
7. Describe the material-removal mechanism when cemented carbides are ground.

Effective Grinding Requirements

8. Name the four factors which a grinder must have for best results when grinding with diamond wheels.
9. Compare the horsepower requirements of a diamond wheel grinding cemented carbide with a CBN wheel grinding high-speed steel.

Wheel Selection

10. What might happen if a wheel with the wrong diamond abrasive or the wrong bond is selected?

11. Explain how the following wheel factors affect the wheel performance.
    a. concentration
    b. wheel diameter
    c. diamond depth
    d. grit size

Wheel Mounting and Preparation

12. What might result if a wheel with excessive runout is used for grinding?
13. What should be the maximum runout for a diamond wheel face; a wheel periphery?
14. List three methods of truing a diamond wheel.
15. What type of tool is used to dress a diamond wheel?

Using Diamond Wheels

16. Why is wheel speed important to the grinding performance?
17. What is the best work speed?
18. If a 100-grit wheel is used, what is the recommended maximum depth of cut?

Carbide Grinding Economics

19. Where do cemented carbides get their hardness and toughness?
20. How is grinding ratio calculated?
21. Name the three forces which act on a wheel in tool and cutter grinding cemented carbides.

Tool and Cutter Grinding Guidelines

22. Name four guidelines which should be considered for best results in regrinding cutting tools.

CHAPTER NINE

Manufacture of Polycrystalline Superabrasives

Chapter 2 covered the technology of manufacturing loose abrasive for use in grinding wheels. The following chapters then expanded on how to use these superabrasive wheels effectively. The remainder of this text will use the same approach for tool blanks and inserts made of polycrystalline diamond (PCD) and polycrystalline cubic boron nitride (PCBN). This chapter covers how these tool blanks are made and their physical properties. Chapters 10 and 11 will expand on how to use them.

The birth of a new technology took place in 1957 when General Electric introduced manufactured diamond. In 1970, polycrystalline superabrasive cutting tools were introduced to cut hard, abrasive materials. These tools consisted of a layer of PCD or PCBN which was bonded to a cemented carbide substrate. Because of their excellent abrasion resistance and long-wearing cutting edges, polycrystalline tools soon became accepted by industry as highly efficient cutting tools.

OBJECTIVES

After completing this chapter you should be able to:

1. Understand the manufacture of polycrystalline superabrasives
2. Be familiar with the properties of polycrystalline tool blanks and inserts and how they differ from carbides, ceramics. and tool steels

THE MANUFACTURING PROCESS

The manufacturing processes for making both diamond and CBN polycrystalline tools are very similar. The abrasive layer starts as finely divided diamond or CBN powder ranging from 2 to 20 μm average size, depending on the ultimate application for the tool. It is sometimes mixed with a small amount of metal. This powder is placed on top of a flat cylindrical carbide substrate from ½ to 1⅜ in. in diameter which is inside a container to hold its shape. One or more of these containers is placed into the high-pressure high-temperature apparatus. The whole assembly is then subjected to about 1 million pounds per square inch (lb/in.^{2}) and over 2192°F (1200°C) and held there for an appropriate time, cooled, and returned to room conditions (Fig. 9-1).

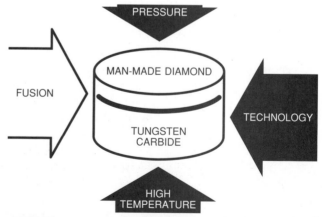

Fig. 9-1 Polycrystalline tool blanks and inserts consist of a thin layer of diamond or CBN bonded to a cemented tungsten carbide substrate (base). *(Courtesy of GE Superabrasives)*

During the several minutes the abrasive powder and substrate are at high pressure and temperature, *sintering* takes place. When powdered materials are compacted and heated to high temperature, they begin to form bonds or bridges from one particle to the next. This reduces the total surface area of the powder and the porosity. Almost all materials do this when the temperature and/or pressure is high enough. This process is called *sintering*.

Polycrystalline tools use a faster type of sintering called *liquid-phase sintering*. The molten metal fills the pores and speeds bond formation between particles. When the cycle is complete, all abrasive particles are tightly bonded to their neighbors. This makes a very tough abrasive layer of approximately 0.020 in. (0.5 mm) thick which is also integrally bonded to the tungsten carbide substrate. The abrasive layer consists of sintered diamond or CBN particles with metal filling all the porosity. The substrate is a cemented tungsten carbide of a type similar to that used for cutting-tool applications.

The rough pressed round tool blanks are then finished to specific shapes and sizes. This is done by various cutting, grinding, lapping, and polishing techniques. Polycrystalline tool blanks are available in a wide variety of shapes and sizes to suit many machining applications (Fig. 9-2).

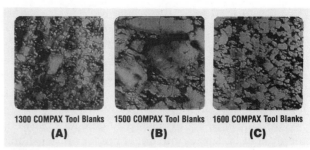

Fig. 9-3 The differences in diamond particle size of the three series of PCD tool blanks. (A) Coarse crystals. (B) Medium-fine crystals. (C) Fine crystals. *(Courtesy of GE Superabrasives)*

CHARACTERISTICS OF POLYCRYSTALLINE DIAMOND TOOLS

The diamond crystals in the polycrystalline superabrasive layer of a Compax tool blank are randomly oriented. This is important because the hardness and abrasion resistance of a diamond crystal change greatly, depending on which crystal face is being tested. By randomly orienting the particles, one aver-

Fig. 9-2 Polycrystalline tool blanks and inserts are available in a wide range of shapes and sizes. *(Courtesy of GE Superabrasives)*

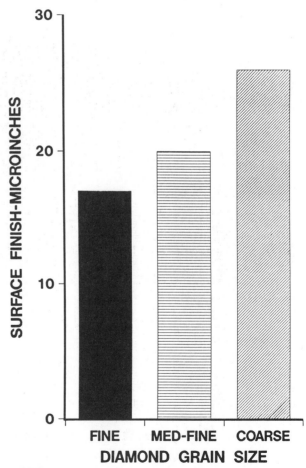

Fig. 9-4 The effects of diamond microstructure on the surface finish produced. *(Courtesy of GE Superabrasives)*

ages the properties so the whole tool surface is uniform. The two most important properties of a polycrystalline tool are hardness and abrasion resistance. These tools are nearly as hard as single-crystal diamond and two to three times as hard as tungsten carbide. The abrasion resistance is equivalent to that measured on a single-point diamond tool.

The abrasive layer by itself would be too costly and too difficult to braze onto a tool holder or tool shank. For this reason and many others, PCD tools use a substrate made of cemented tungsten carbide. This provides mechanical support for the abrasive layer and easily brazes to the toolholder.

Polycrystalline diamond tools are manufactured in three grades and a wide variety of shapes and sizes. The particle size distribution of diamond used in the abrasive layer determines the grade. The different distributions are illustrated in Fig. 9-3. The coarsest grade is the 1500 series. These tools have the highest impact resistance. The finest is the 1600 series, which has the ability to form a very sharp edge. Intermediate in grain size between the two is the most widely used 1300 series.

While all of these grades have the same hardness characteristics, they vary greatly in abrasion resistance and their ability to produce surface finish. This ability to vary the properties of PCD means that a wider range of applications which require the use of a PCD tool can be satisfied. The effect of microstructure of the toolbit on surface finish of the workpiece is shown in Fig. 9-4. In the machining of soft metals such as copper and pure aluminum, and wood fiber products, a high degree of abrasion resistance is not required but a good surface finish is important. Thus the fine-grain 1600 grade product satisfies this combination of requirements. At the other extreme, the machining of new silicon and silicon carbide aluminum composites require extraordinary abrasion resistance and the coarse grain 1500 grade would be needed. Relative abrasion resistances of the various microstructures are shown in Fig. 9-5.

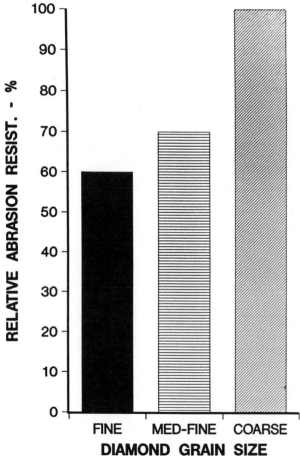

Fig. 9-5 The effects of diamond microstructure on the abrasion resistance of PCD tools. *(Courtesy of GE Superabrasives)*

Table 9-1 COMPAX DIAMOND TOOL BLANKS

SEGMENTS

Product Number	ℓ°	d	ℓ	t
1311	45	—	3.8	1.5
1315	60	—	3.8	3.2
1316	60	—	3.8	1.5
1320	90	—	3.9	3.2
1321	90	—	3.9	1.5
1325	180	—	8.1	3.2
1326	180	—	8.1	1.5
1330	360	8.1	—	3.2
1331	360	8.1	—	1.5
1511	45	—	6.3	1.5
1515	60	—	6.4	3.2
1516	60	—	6.4	1.5
1520	90	—	6.5	3.2
1521	90	—	6.5	1.5
1525	180	—	13.2	3.2
1526	180	—	13.2	1.5
1530	360	13.2	—	3.2
1531	360	13.2	—	1.5
1611	45	—	6.3	1.5
1615	60	—	6.4	3.2
1616	60	—	6.4	1.5
1620	90	—	6.5	3.2
1621	90	—	6.5	1.5
1625	180	—	13.2	3.2
1626	180	—	13.2	1.5
1630	360	13.2	—	3.2
1631	360	13.2	—	1.5

TRIANGLES

Product Number	ℓ°	d	ℓ	t	w
60T5/1.5 1300	60	—	5.0	1.5	—
90T5/1.5 1300	90	—	5.0	1.5	—
60T5/1.5 1500	60	—	5.0	1.5	—
90T5/1.5 1500	90	—	5.0	1.5	—
90T7.5/1 1500	90	—	7.5	1.5	—

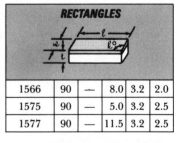

RECTANGLES

Product Number	ℓ°	d	ℓ	t	w
1566	90	—	8.0	3.2	2.0
1575	90	—	5.0	3.2	2.5
1577	90	—	11.5	3.2	2.5

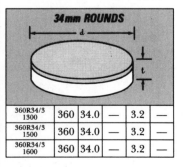

34mm ROUNDS

Product Number	ℓ°	d	ℓ	t	
360R34/3 1300	360	34.0	—	3.2	—
360R34/3 1500	360	34.0	—	3.2	—
360R34/3 1600	360	34.0	—	3.2	—

(Nominal diamond layer thickness is 0.5 mm)

(Courtesy of GE Superabrasives.)

The range of sizes of segments, triangles, rectangles, and round blanks which are available for tool fabrication are shown in Table 9-1.

TYPES OF MATERIALS CUT

Polycrystalline diamond tools are used for machining applications of nonferrous or nonmetallic materials, primarily where the workpiece is abrasive. These materials are generally considered difficult to machine due to their abrasive character.

The largest group of nonferrous metals is typically soft but have hard particles dispersed in them; for example, silicon suspended in aluminum or glass fibers in plastic. It is the hard abrasive particle that destroys the cutting edge of conventional tools. Diamond is harder than the abrasive particle and tends to shear the hard particle rather than pushing it out of the way or dulling the cutting edge. PCD blank tools may often reach a wear life of 100 times that of cemented carbide tools in such an abrasive machining application.

A growing new category of nonmetallic materials are the ceramics and composites. These materials are both hard and abrasive and rely on the hardness of the diamond to overcome their own abrasive character. PCD cutting tools are capable of cutting the hard abrasive inclusions in these materials cleanly without the rapid dulling of the cutting edge.

The materials most successfully machined with PCD tools fall into three general categories: metals, nonmetals, and composites. Table 9-2 lists the more common materials which can be cost-effectively machined with PCD tools.

CHARACTERISTICS OF POLYCRYSTALLINE CUBIC BORON NITRIDE TOOLS

The process of manufacturing PCBN blanks produces the same type of beneficial characteristics as in the case of PCD blanks. The random orientation of the CBN grains in the PCBN layer provides hardness, toughness, and abrasion resistance characteristics which can be exceeded only by PCD. In addition, the cemented carbide substrate provides the same level of support and brazability as for PCD.

The most important grade or type of PCBN tool blank structure is known as a *BZN* blank*. In addition to this structure, there is a PCBN composite known as *BZN* 8000* tool blanks. These blanks are manufactured in a process very similar to that by which BZN blanks are made except that a ceramic phase is used in the process. The addition of ceramic improves the PCBN layer's performance and increases its overall wear resistance in certain applications.

*Trademark of General Electric Company, USA.

Table 9-2	**PCD CUTTING TOOL APPLICATIONS**

Nonferrous Metals	Nonmetallic Materials	Composites
Aluminum Alloys	Alumina, Fired	Asbestos
Babbitt Alloys	Bakelite	Fiberglass Epoxy
Brass Alloys	Beryllia	Filled Carbons
Bronze Alloys	Ceramics	Filled Nylon
Copper Alloys	Epoxy	Filled Phenolic
Lead Alloys	Glass	Filled P.V.C.
Manganese Alloys	Graphite	Filled Silica
Silicon-Aluminum	Macor	Filled Teflon
Silver, Platinum	Rubber, Hard	Wood, Manufactured
Tungsten Carbides	Various Plastics	
Zinc Alloys		

(Courtesy of GE Superabrasives.)

Table 9-3 **BZN COMPACTS**

BZN BLANKS

Product Number	Product Shape	Nominal Dimensions (mm)				
		t	d	k	l	ic
6315	▼	3.2	0.6	60°	3.6	—
6320	◗	3.2	0.6	90°	3.8	—
6325	◡	3.2	0.6	180°	8.0	—
6330	●	3.2	0.6	360°	—	8.0
6515	▼	3.2	0.6	60°	6.1	—
6520	◗	3.2	0.6	90°	6.2	—
6525	◡	3.2	0.6	180°	12.7	—
6530	●	3.2	0.6	360°	—	12.7
6730	●	4.8	0.8	360°	—	23.5
6731	●	3.2	0.8	360°	—	23.5

BZN INSERTS

Product Number	Product Shape	Nominal Dimensions (mm)		
		t	d	ic
BRNG-42	●	3.2	0.6	12.7
BRNG-43	●	4.8	0.8	12.7
BRNG-53	●	4.8	0.8	15.9
BSNG-432 BSNG-433 BSNG-434	■	4.8	0.8	12.7
BSNG-532 BSNG-534	■	4.8	0.8	15.9
BTNG-433	▲	4.8	0.8	12.7

(Courtesy of GE Superabrasives.)

BZN 8000 BLANKS

Product Number	Product Shape	Nominal Dimensions (mm)		
		t	k	l
60T 4.5/3 B-8000	▲	3.2	60	4.5
60T 6.5/3 B-8000	▲	3.2	60	6.5
90T 4.5/3 B-8000	◣	3.2	90	4.5
90T 6.5/3 B-8000	◣	3.2	90	6.5
60P 4.0/3 B-8000	▼	3.2	60	4.0
60P 6.0/3 B-8000	▼	3.2	60	6.0
90P 4.0/3 B-8000	◗	3.2	90	4.0
90P 6.0/3 B-8000	◗	3.2	90	6.0
360R 34/3 B-8000	●	3.2	360	34.0

*Nominal Borazon CBN layer thickness is 0.8 mm.

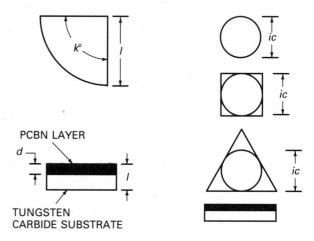

PCBN LAYER

d

TUNGSTEN CARBIDE SUBSTRATE

Polycrystalline cubic boron nitride blanks are available in a range of segments, triangles, rectangles, and rounds, similar to those available for PCD. These are listed in Table 9-3. In addition, full-faced PCBN inserts in rounds, squares, and triangles, made to industry standard dimensions, are available to fit appropriate standard toolholders.

TYPES OF METAL CUT

Polycrystalline cubic boron nitride cutting tools are used on lathes and turning centers for machining round surfaces and on milling machines and machining centers for machining flat surfaces. These cutting tools have been used successfully for straight turning, facing, boring, grooving, profiling, and milling operations. PCBN blank tools and inserts are capable of removing materials at much higher rates than conventional cutting tools, with far longer tool life. This results in a great increase in productivity and lower cost per piece machined.

Polycrystalline cubic boron nitride tool blanks and inserts were designed for hard, difficult-to-cut (DTC) materials and hard abrasive materials which were previously ground because they were considered unmachinable. Wherever PCBN cutting tools were used to replace a grinding operation, machining time was greatly reduced because of the higher metal-removal rate.

The three general areas of metals where PCBN cutting tools have found the best applications and proved to be cost-effective are:

1. **Hardened steels** over 45 Rc hardness. These include hardened steels such as 4340, 8620, M-2, T-15, etc.
2. **Cast irons** ranging from 180 to 240 Brinell hardness, such as pearlitic gray cast iron, Ni-Resist, etc.
3. **Superalloys,** such as high-nickel alloys used in the aerospace industry for jet engine parts.

Table 9-4 illustrates some of the more common metals which are machined efficiently with PCBN cutting tools.

The best applications for PCBN cutting tools are on materials where conventional cutting-tool edges of cemented carbides and ceramics are breaking down too quickly. PCBN tools are especially important on expensive machining systems such as numerical-control (NC) machine tools and flexible manufacturing systems. Their long-lasting cutting edges are capable of transferring the accuracy of computer-controlled machine tools or systems to the workpiece, thereby producing accurate parts, increasing productivity, and reducing expensive machine downtime on high-investment equipment.

PROPERTY COMPARISON WITH OTHER TOOL MATERIALS

A brief summary of the physical properties of PCD and PCBN in comparison with other selected tool materials is found in Tables 9-5A and 9-5B. It is apparent that the properties of PCD and PCBN are indeed "super" in comparison with other tool tip materials. The superiority in the cases of hardness and thermal conductivity are clear from the values shown. In the case of the transverse rupture strength values, it should be remembered that the PCD and PCBN layers are bonded to a cemented carbide substrate and the resulting properties are superior to any other tool material.

This comparison shows that PCD and PCBN are the hardest, most thermally conductive, and by far the most abrasion-resistant tool materials available to the metalworking industries worldwide.

Of particular importance in comparing properties of cutting-tool materials is the so-called hot-strength property. This characteristic will determine the upper limit of cutting speed to which a tool tip can be subjected in the machining process. The higher the hot strength, the higher the allowable cutting speed, and the higher the productivity of the machining operation. Figure 9-6 illustrates the comparative strengths of several tool materials as a function of temperature. Again the properties of PCD and PCBN are super in comparison with all other materials.

Table 9-4	PCBN TOOL APPLICATIONS

Hard cast iron
  Ni-Hard
  Alloy cast iron
  Chilled cast iron
  Nodular cast iron

Soft cast iron
  Gray cast iron

Sintered iron
  Powdered metal
  Sintered iron

Hardened steels
  Tool steels
  Die steels
  Case hardened steels
  A, D, M series steels
  Bearing steels

Superalloy
  Inconel-718, -901, -600
  Rene 77, 95
  Hastalloy
  Waspalloy
  Stellite
  Colmonoy
  K-Monel

Table 9-5A **COMPARISON OF SELECTED PROPERTIES OF CUTTING-TOOL MATERIALS**

Comparison of Selected Properties of Cutting Tool Materials

	Knoop Hardness, kg/mm²	Transverse Rupture Strength, lb/in.²	Thermal Conductivity, W/(m·K)
Polycrystalline diamond	6000	140,000–200,000	500
Polycrystalline CBN	3500	100,000	100
Cemented carbides	1500–1800	250,000–300,000	40–80
Silicon nitride	1700	70,000–110,000	15–35
Aluminum oxide	1600	40,000–50,000	14–17
High-speed steel	700–1000	350,000–400,000	40–80

(Courtesy of GE Superabrasives.)

Table 9-5B **CUTTING-TOOL SELECTION**

Resistance to abrasion				
1	5–10	10–30	50	100
Carbide	Coated carbides	Ceramics	cubic boron nitride	diamond

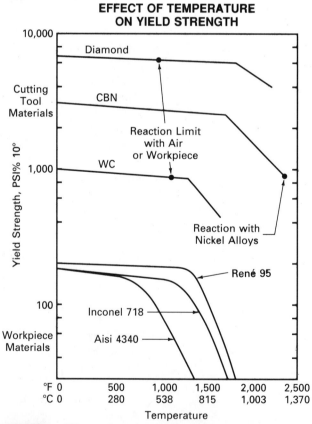

EFFECT OF TEMPERATURE ON YIELD STRENGTH

Fig. 9-6 Effects of temperature on yield strength. *(Courtesy of GE Superabrasives)*

1. Describe the composition of polycrystalline cutting tools.
2. Why are they finding wide use in industry?

The Manufacturing Process

3. List the manufacturing process for polycrystalline tools.

Characteristics of PCD Tools

4. Why are the properties of a PCD tool blank surface uniform?
5. What purpose does the cemented carbide substrate serve?
6. What effects does the microstructure of the diamond layer have on the surface finish produced?

7. What effects does the microstructure of the diamond layer have on the abrasion resistance?

Types of Materials Cut

8. Name the three classes of materials which should be machined with PCD tools.

Characteristics of PCBN Tools

9. What properties does the random orientation of the grains in the PCBN layer provide?
10. How does the BZN 8000 structure differ from other grades?

Types of Metal Cut

11. Name the five types of metals commonly machined with PCBN tools.

CHAPTER TEN

Polycrystalline Cubic Boron Nitride Tools

The development of polycrystalline cubic boron nitride (PCBN) cutting tools has allowed material-removal rates for difficult-to-machine materials to be greatly increased. The superhardness of CBN, especially at high temperature, combined with the toughness and strength of the tungsten carbide base, allows the machining of hardened steels, abrasive cast irons, and tough high-temperature alloys at high removal rates with long tool life even when heavy or interrupted cuts are taken. The polycrystalline form of CBN features nondirectional, consistent properties that resist chipping and cracking, and provide uniform hardness and abrasion resistance in all directions. Tables 10-1 and 10-2 list the properties of the polycrystalline CBN layer and the cemented tungsten carbide substrate (base).

OBJECTIVES

After completing this chapter you should be able to:

1. Understand the properties of PCBN tool blanks and inserts and how they differ from carbides and ceramics

2. Select the proper PCBN tools for the machining of various metals

3. Set up the cutting tool and machine for cutting with PCBN tools

4. Understand the conditions necessary to achieve maximum productivity from PCBN cutting tools

Table 10-1 **PROPERTIES OF POLYCRYSTALLINE CBN LAYER**

Transverse rupture strength	105×10^3 lb/in.^{2}, 724 MPa
Modulus of elasticity	125×10^6 lb/in.^{2}, 862 MPa
Knoop hardness	3500 kg/mm^2
Thermal conductivity	1-2 W/(cm $\cdot$ K)

Table 10-2 **TYPICAL PROPERTIES OF CEMENTED TUNGSTEN CARBIDE SUBSTRATE**

Transverse rupture strength	390×10^3 lb/in.^{2}, 2700 MPa
Modulus of elasticity	80×10^6 lb/in.^{2}, 551 160 MPa
Knoop hardness	1200 kg/mm^2
Thermal conductivity	1 W/(cm $\cdot$ K)

TYPES AND SIZES OF PCBN TOOLS

Polycrystalline cubic boron nitride tool blanks and inserts are a combination of a layer of CBN bonded to a cemented tungsten carbide substrate. A typical blank is illustrated in Fig. 10-1. After the sintering operation, the rough-pressed round tool blanks are finished to specific shapes and sizes to suit various machining applications (Fig. 10-2).

Finished ready-to-use PCBN tools are available from tool suppliers and fall into three categories: tipped inserts, inserts, and brazed-shank tools (Fig. 10-3).

Tipped inserts (Fig. 10-3A) are available in most regular carbide insert shapes and are generally the most economical to purchase. These inserts are manufactured by machining a pocket in the carbide and brazing a PCBN blank in place. They have the same wear life per cutting edge as a PCBN insert; however, they have only one cutting edge. When this cutting edge becomes dull, it is necessary to regrind the tool. Reground cutting edges have the same machining life as a new cutting tool.

PCBN inserts (Fig. 10-3B) consist of a layer of PCBN bonded to a cemented carbide substrate. PCBN inserts are available as squares, rounds, and triangles and are generally very cost-effective since the tool manufacturer can downsize them repeat-

edly, providing new cutting edges. The coding system used for PCBN inserts is shown in Table 10-3.

Brazed-shank tools (Fig. 10-3C) are available in most common cutting-tool forms. They are made by machining a pocket in the proper-style tool shank and brazing a PCBN blank in place. They can be specially ordered from the manufacturer to suit a wide variety of machining applications.

PROPERTIES OF POLYCRYSTALLINE CUBIC BORON NITRIDE

Cubic boron nitride is a manufactured tool material which is not found in nature. PCBN cutting tools (see Table 10-4) contain properties which cutting tools must have to cut extremely hard or abrasive materials at high metal-removal rates. These properties are hardness at high temperature, abrasion resistance, high strength, high impact resistance, and thermal conductivity. There are two types of PCBN tool blanks, each with different properties—basic PCBN and PCBN composites. The properties of various cutting-tool materials are compared in Tables 9-5A, 9-5B, and Fig. 9-6.

Basic Polycrystalline Cubic Boron Nitride Tool Blanks

The *hardness* of the polycrystalline CBN layer is far higher than that of cemented tungsten carbide. Because the crystals of CBN are randomly oriented, hardness and *abrasion resistance* are uniform in all directions.

The CBN crystal bonding imparts uniformly *high strength* to the polycrystalline CBN layer. The tungsten carbide substrate provides *high impact resistance* needed to withstand the shock of severe interrupted cuts.

The basic PCBN layer has a *high thermal conductivity*, higher than that of carbides and ceramics. This allows greater cutting-tool heat dissipation (transfer) in machining applications than that seen with either carbide or ceramic tool materials.

When cutting at high material-removal rates, cutting temperatures may be so high that they weaken or soften conventional cutting-tool materials. This is not the case with PCBN tools. The polycrystalline CBN cutting-tool edge keeps its strength and hardness at machining temperatures of up to 2190°F (1200°C). It also resists oxidation at temperatures up to 1200°C, and its chemical reaction with metals—including iron, nickel, and cobalt—at this temperature is relatively small.

Polycrystalline Cubic Boron Nitride Composites

The characteristics of PCBN composite tools are somewhat different than the basic PCBN tools. Since

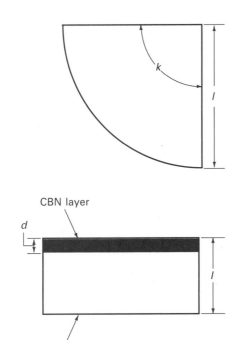

CBN layer

Tungsten carbide substrate

Fig. 10-1 PCBN tool blanks and inserts consist of a thin layer of CBN bonded to a cemented tungsten carbide substrate (base). *(Courtesy of GE Superabrasives)*

BZN BLANKS

Product Number	Product Shape	Nominal Dimensions (mm)				
		t	d	k	l	ic
6315		3.2	0.6	60°	3.6	—
6320		3.2	0.6	90°	3.8	—
6325		3.2	0.6	180°	8.0	—
6330		3.2	0.6	360°	—	8.0
6515		3.2	0.6	60°	6.1	—
6520		3.2	0.6	90°	6.2	—
6525		3.2	0.6	180°	12.7	—
6530		3.2	0.6	360°	—	12.7
6730		4.8	0.8	360°	—	23.5
6731		3.2	0.8	360°	—	23.5

BZN INSERTS

Product Number	Product Shape	Nominal Dimensions (mm)		
		t	d	ic
BRNG-42		3.2	0.6	12.7
BRNG-43		4.8	0.8	12.7
BRNG-53		4.8	0.8	15.9
BSNG-432 BSNG-433 BSNG-434		4.8	0.8	12.7
BSNG-532 BSNG-534		4.8	0.8	15.9
BTNG-433		4.8	0.8	12.7

BZN 8000 BLANKS

Product Number	Product Shape	Nominal Dimensions (mm)		
		t	k	l
60T 4.5/3 B-8000		3.2	60	4.5
60T 6.5/3 B-8000		3.2	60	6.5
90T 4.5/3 B-8000		3.2	90	4.5
90T 6.5/3 B-8000		3.2	90	6.5
60P 4.0/3 B-8000		3.2	60	4.0
60P 6.0/3 B-8000		3.2	60	6.0
90P 4.0/3 B-8000		3.2	90	4.0
90P 6.0/3 B-8000		3.2	90	6.0
360R 34/3 B-8000		3.2	360	34.0

*Nominal Borazon CBN layer thickness is 0.8 mm.

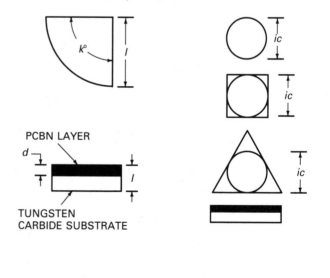

Fig. 10-2 Polycrystalline CBN tool blanks and inserts are available in a wide range of sizes and shapes. (*Courtesy of GE Superabrasives*)

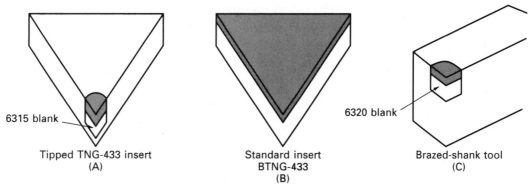

PCBN layer ■

| 6315 blank | 6320 blank |
| Tipped TNG-433 insert (A) | Standard insert BTNG-433 (B) | Brazed-shank tool (C) |

Fig. 10-3 The three common categories of PCBN cutting tools. (A) Tipped inserts. (B) Full-faced inserts. (C) Brazed shank. *(Courtesy of GE Superabrasives)*

about half of the volume of the layer is ceramic, the PCBN composites have the following characteristics as compared to the basic PCBN:

1. The wear resistance is higher on certain finishing applications because of its superior chemical and thermal properties.
2. The impact resistance is less and therefore they are not generally recommended for interrupted cuts.

PCBN SELECTION

A general comparison of the properties of PCBN and PCBN composite is shown in Table 10-5A. The basic PCBN blank has the combination of properties which will make it the most widely used PCBN for machining applications, but the superior chemical and thermal stability properties of the PCBN composite give it higher wear resistance for certain hardened-steel finishing applications.

Table 10-3 PCBN INSERT CODING SYSTEM

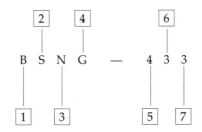

1 B—PCBN blank

2 S—square   R—round   T—triangle

3 N—negative rake   P—positive rake
      (specify angle, 6° standard)

4 G—precision insert tolerances
    U—utility insert tolerances (not available)

5 number of 1/8-in. IC (i.e., 4/8 = 1/2-in. IC square) ◎ 1/2-in.

6 number of 1/16-in. thickness (i.e., 3/16 in. thick)

7 number of 1/64-in. nose radius (i.e., 3/64 nose radius applicable for squares and triangles only)

Table 10-4   PCBN TOOL APPLICATIONS

Hard cast iron
  NI-Hard
  Alloy cast iron
  Chilled cast iron
  Nodular cast iron

Soft cast iron
  Gray cast iron

Sintered iron
  Powdered metal
  Sintered iron

Hardened steels
  Tool steels
  Die steels
  Case hardened steels
  A, D, M series steels
  Bearing steels

Superalloy
  Inconel-718, -901, -600
  Rene 77, 95
  Hastalloy
  Waspalloy
  Stellite
  Colmonoy
  K-Monel

The general guidelines for selecting PCBN or PCBN composite tools are shown in Table 10-5B. The main factors for selection are the properties of the workpiece and the type of the machining operation (roughing or finishing).

A more detailed set of guidelines for the particular case of hardened steels is shown in Table 10-5C. This table shows the advantages of each type of PCBN on the basis of their specific properties. PCBN is tough enough to handle surface scale and out-of-round workpiece surfaces. The PCBN composite can handle higher work speeds but is not suitable for rough uneven workpiece surfaces.

ADVANTAGES OF PCBN CUTTING TOOLS

The advantages that PCBN cutting tools offer the metalworking industry more than offset their higher

Table 10-5A   PROPERTY COMPARISON

	PCBN	PCBN Composites
Hardness	Excellent	Excellent
Toughness	Good	Fair
Chemical stability	Good	Very good
Thermal stability	Good	Very good

Table 10-5B   PCBN SELECTION GUIDELINES

Material Type	Machining Operation	
	Roughing	Finishing
Hardened steel (>45 Rc)	PCBN	PCBN composite
Gray cast iron (200 BHN)	PCBN	PCBN
Hard facing alloys		PCBN
Superalloys (>35 Rc)	PCBN	PCBN

Table 10-5C   PCBN SELECTION FOR MACHINING HARDENED STEELS (>45 RC) AND HARD FACING METALS

Select	Workpiece surface condition	Speed, sf/min (m/min)	Feed rate, in./rev (mm/rev)	Depth of cut, in. (mm)

PCBN	Can have:
	  Scale
	  Out-of-round
		230–350
		(70–106)
			0.004–0.020
			(0.10–0.50)
				0.005–0.150
				(0.12–3.80)

PCBN composite	Must be:
	  Clean, no scale
	  Uniform depth
	    of cut
		330–430
		(100–130)
			0.004–0.008
			(0.10–0.20)
				0.004–0.030
				(0.10–0.76)

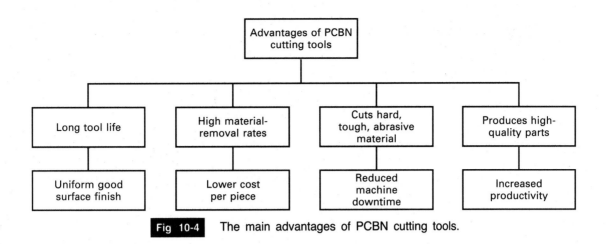

The main advantages of PCBN cutting tools.

initial costs. Primarily designed to machine very hard and difficult-to-grind (DTG) metals, PCBN tools are capable of greatly improving productivity, reducing scrap parts, and increasing the quality of the product.

Let us examine the advantages of PCBN superabrasive cutting tools which can be used for turning and milling operations (Fig. 10-4).

Long Tool Life. PCBN cutting tools have properties which resist chipping and cracking and provide uniform hardness and abrasion resistance in all directions. They may outperform conventional cutting tools by as much as 50 times. Reduced tool wear results in closer tolerances on workpieces, and fewer tool adjustments keep machine downtime to a minimum. The increased reliability of PCBN cutting tools make them a major factor in automating machining operations.

High Material-Removal Rates. Because PCBN cutting tools are so hard and resist abrasion so well, cutting speeds in the range of 250 to 400 ft/min are possible on hardened steels and speeds of 5000 ft/min are possible in pearlitic cast iron. PCBN cutting tools can withstand the high temperatures created by increased speeds and feeds. This results in higher material-removal rates with less tool wear, which reduces the total machining cost per piece.

Cutting Hard, Tough Materials. PCBN cutting tools are so hard that they can efficiently machine steel with a hardness of Rc 45 and above. PCBN is also used for machining cobalt-base and nickel-base high-temperature alloys with a hardness of Rc 35 and above. Many of these materials are so hard and tough that they must be machined at relatively low speeds. Conventional grinding is also used but is a relatively slow metal-removal process. PCBN cutting tools have a high resistance to impact and therefore can machine parts with interrupted cuts and tough abrasive scale.

High-Quality Products. Because the cutting edges of PCBN cutting tools wear very slowly, they produce high-quality parts faster and at a lower cost per piece than do conventional cutting tools. The need for the inspection of parts is greatly reduced, as is the adjustment of the machine tool to compensate for cutting-tool wear or maintenance. This results in better control over workpiece shape and size, which produces consistent part quality far beyond that possible with other cutting tools.

Uniform Surface Finish. Surface finishes of less than 10 μin. are possible, which often eliminate the need for relatively slow finishing operations such as conventional grinding.

Lower Cost per Piece. PCBN cutting tools stay sharp and cut efficiently through long production runs. This results in better control over workpiece shape and size and fewer cutting-tool changes. This lowers manufacturing costs per piece by reducing inspection time and increasing machine uptime.

MACHINE-TOOL CONDITIONS

Experience has shown that production machine tools used for machining hardened steels or superalloys with cemented tungsten carbide and ceramic inserts are usually suitable for machining operations with PCBN blank tools. This does not apply to the machining of gray cast iron, as much higher speed ranges are required to cut this material effectively and economically.

It is recommended that any machine on which PCBN will be used be checked beforehand for worn bearings, loose gibs, and excessive wear on other machine components. Workholding fixtures should also be examined for rigidity before undertaking normal facing, turning, boring, and similar operations.

Newer machines selected for use with PCBN tools

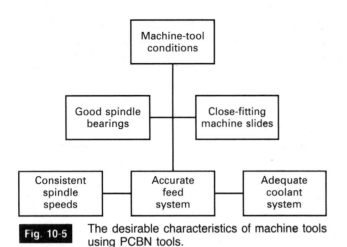

Fig. 10-5 The desirable characteristics of machine tools using PCBN tools.

should be equipped with high dynamic stiffness and static rigidity with adequate horsepower and be capable of cutting at high speeds. These features can take advantage of the full potential that PCBN blank tools and inserts offer—high material removal at increased speeds and feeds. This is especially true of automated machine tools.

To work effectively, PCBN cutting tools *must be used* on machines which have the following characteristics (Fig. 10-5):

Spindle Bearings. Loose spindle bearings will cause vibration and chatter during the machining operation, shortening the cutting tool life and producing a poor surface finish and inaccurate work.

Machine Slides. The slides must be close-fitting to prevent vibration and chatter, which would reduce the effectiveness of superabrasive cutting tools and result in poor tool life, poor surface finish, and inaccurate work.

Spindle Speeds. High spindle speeds and the ability to handle the torque required for high metal-removal rates are necessary to keep the superabrasive cutting tool operating at its best efficiency.

INCREASING EDGE STRENGTH

Fig. 10-6 Common PCBN insert shapes and their relative edge strengths. *(Courtesy of GE Superabrasives)*

Variations in spindle speeds reduce the efficiency of the cutting action and shorten the life of the PCBN cutting tool.

Feeds. A reliable feed system is required in order to produce good surface finishes and maintain good cutting tool life.

Coolant System. See page 169 for a detailed discussion on the use of cutting fluids.

CUTTING-TOOL CHARACTERISTICS

As with all types of cutting tools, they must be properly prepared and used under the conditions which will provide the highest machining performance. To maintain good control over part quality and finish, it is always wise to establish the life or cutoff point of a cutting edge (usually after a certain number of pieces are cut) and make a practice of changing cutting tools at this point. PCBN cutting tools, if run until they are excessively worn, will cut inefficiently and increase machining temperatures and the cutting force required. This, in turn, will produce poor surface finishes and affect part accuracy and geometry. Badly worn PCBN tools will more likely suffer impact damage, especially during interrupted cuts, because the shock force of a dull tool is much higher than that of a sharp tool.

NOTE: Do not run PCBN tools beyond normal wear limits. Excessively worn tools may not be economically reground, resulting in increased tool costs.

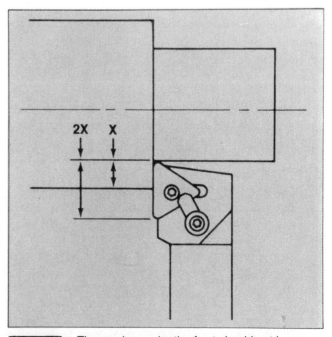

Fig. 10-7 The maximum depth of cut should not be more than half the length of the cutting edge. *(Courtesy of Carboloy Inc.)*

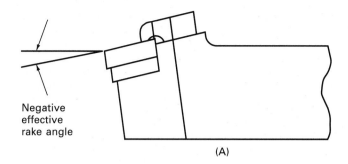

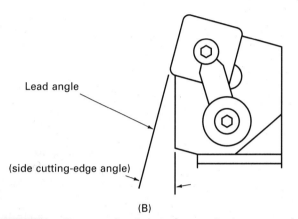

Negative
effective
rake angle

(A)

Lead angle

(side cutting-edge angle)

(B)

Fig. 10-8 Suggested rake angles and clearances for PCBN cutting tools. (A) Negative rake. (B) Lead angle (no less than 15°). *(Courtesy of Carboloy Inc.)*

Three important factors regarding the selection and preparation of a PCBN cutting tool for machining should be considered—overall tool shape, tool geometry, and edge preparation.

Overall Tool Shape

The strength of any cutting-tool edge is affected by the basic shape of the tool. Figure 10-6 shows the effect that tool shape has on the edge strength. As the overall width and depth of the tool shape is improved from a triangle on the right to a circle on the left, the overall strength of the tool is increased. Always select

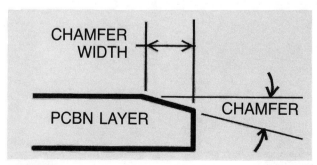

CHAMFER
WIDTH

PCBN LAYER

CHAMFER

Fig. 10-9 The recommended edge preparation for PCBN cutting tools. *(Courtesy of GE Superabrasives)*

the strongest tool shape which can be used to machine the workpiece to the form required (Fig. 10-7).

Round Inserts. These inserts generally provide more cutting edges per side than any other insert shape. The number of cutting edges will depend on the depth of cut and the care taken in indexing. Round inserts are by far the strongest insert shape because they do not have corners; however, they do have some important disadvantages.

They can be used only for machining workpieces which do not require sharp corners. Also, the biggest depth of cut which can be taken with round inserts is half the insert diameter.

Square Inserts. Although square inserts have fewer cutting edges and less corner strength than round inserts, they are the ones most widely used. They can be set to almost any lead angle and produce a wide variety of corner shapes.

Triangular Inserts. These inserts are possibly the most versatile of insert shapes because they can be used to produce a wide range of shapes or forms on a workpiece. For any given size, triangular inserts have longer cutting edges than do any other insert shape; however, they are weaker.

Tool Geometry

The geometry (shape) of the cutting tool and how it meets the workpiece are very important factors which determine how well and how long it will cut efficiently. To obtain the longest tool life, efficient metal-removal rates, and acceptable surface finishes, the following points should be observed:

1. Always use negative-rake tools wherever possible (Fig. 10-8A). Negative-rake tools can withstand higher cutting forces (Fig. 10-8B).
2. The lead or side cutting edge angle should be as large as possible and rarely less than 15°.
   a. A large lead angle spreads the cut over a wide section of the cutting edge, resulting in a thinner chip, which reduces loading on the tool blank or insert.
   b. The feed per revolution can be increased without increasing the chances of the cutting edge chipping due to excessive loading.
   c. A large lead angle helps to reduce notching at the depth-of-cut line, which is the result of the scale or hardness of a workpiece.

Edge Preparation

The edge preparation of any cutting tool is important to the life and efficiency of the cutting action, especially when machining hard, abrasive materials. If close attention is given to the preparation of the cut-

Table 10-6   EDGE PREPARATION GUIDELINES

Workpiece Material	Operation	Superabrasive	Hone Edge	Chamfer Dimensions
Hardened steel	Roughing	PCBN	No	15° × 0.008 in. (0.20 mm)
Hardened steel	Finishing	PCBN composite	No	20° × 0.004 in. (0.10 mm)
Soft cast iron	Roughing	PCBN	No	15° × 0.008 in. (0.20 mm)
Soft cast iron	Finishing	PCBN	0.001 in. (0.02 mm) max or	15° × 0.008 in. (0.20 mm)
Hard facing alloys	All	PCBN	No	15° × 0.008 in. (0.20 mm)
Superalloys	Roughing	PCBN	No	15° × 0.008 in. (0.20 mm)
Superalloys	Finishing	PCBN	0.001–0.003 in. (0.02–0.07 mm)	None

ting edge, the life of the cutting tool will be extended, thereby increasing the productivity of the cutting operation. The following points must be observed when preparing the edge of PCBN and PCBN composite cutting tools.

Sharp corners on cutting tools act as stress concentration points and can cause premature tool failure. Two methods are available to overcome this problem:

1. Honing a radius on the edge.
2. Chamfering the cutting edge.

Depending on the type of the cutting operation, one or both of these methods may be required. The principles of chamfering and its dimensions are shown in Fig. 10-9. The guidelines for applying honing and chamfering edge preparations are shown in Table 10-6.

Honing should be carefully done with a very fine grit diamond honing tool. Where a chamfer is required, it must be ordered and specified from the tool supplier, as few metalworking operations will be equipped to put this precise edge preparation on PCBN tools.

TOOL

| Fig. 10-10 | The crystal structure of a metal deforms just ahead of the cutting tool during a machining operation. *(Courtesy of Cincinnati Milacron Inc.)* |

| Fig. 10-11 | When brittle metals are cut, discontinuous or segmented chips are produced. *(Courtesy of Cincinnati Milacron Inc.)* |

Nose Radius

The selection of the proper nose radius for a PCBN tool is as important as in the case of conventional tool materials. A nose radius as large as possible should always be used because:

1. It produces a better surface finish.
2. A larger tool contact with the workpiece provides better distribution of heat. This will allow higher, more productive cutting speeds.
3. The larger the radius, the stronger will be the cutting-tool point.

PRINCIPLES OF METAL CUTTING

Metal cutting is a process where a wedge-shaped cutting tool contacts the workpiece to remove a layer of material in the form of a chip.

Stresses are created during metal-cutting operations and have the following results:

1. Because of the forces exerted by the cutting tool edge, a compression of metal occurs in the material.
2. As the cutting tool moves forward during the cut, or as the workpiece revolves, stress lines concentrate at the cutting-tool edge and radiate into the workpiece (Fig. 10-10).
3. This concentration of stresses causes the chip to shear from the workpiece and flow along the chip-tool interface.
4. By either plastic flow or rupture, the metal tries to flow along the chip-tool interface. Since most metals are ductile to some degree, a plastic flow generally occurs.

Whether plastic flow or rupture occurs as the metal flows along the chip-tool interface will determine the type of chip which will be produced. When brittle metals such as cast iron are being cut, the metal has a tendency to rupture and produce discontinuous or segmented chips (Fig. 10-11). When relatively ductile metals are being cut, a plastic flow occurs and continuous or flow-type chips are produced (Fig. 10-12).

Effect of Metal on Chip Formation and Tool Wear

The type of metal being cut will have a great effect on the type of chip produced and the amount of forces and power required for the metal-removal operation. Steels, superalloys, and cast irons will all produce different chips and have a different effect on the wear rate of the cutting tool.

Tool Life

The wear or abrasion of the cutting tool will determine its life. Three types of wear are generally associated with cutting tools: *flank wear*, *nose wear*, and *crater wear* (Fig. 10-13).

Flank wear (Fig. 10-13A) is the wear occurring on the side of the cutting edge as a result of friction between the side of the cutting-tool edge and the metal being machined. Too much flank wear increases friction and makes more power necessary for machining. When the flank wear becomes excessive, the tool requires regrinding.

Nose wear (Fig. 10-13B) is the wear occurring on the nose or point of the cutting tool as a result of friction between the nose and the metal being machined. Wear on the nose of the cutting tool affects the quality of the surface finish on the workpiece.

Crater wear (Fig. 10-13C) is the wear occurring a slight distance away from the cutting edge as a result of the chips sliding along the chip-tool interface due to a built-up edge on the cutting tool. Too much crater wear eventually breaks down the cutting edge. Other factors which can cause tool wear or breakdown are chipping, thermal cracking, and fracture.

Chipping (Fig. 10-13D) occurs when very small pieces of the cutting edge are broken off and not worn away. This occurs when the force of the cutting operation becomes greater than the strength of the small sections of the cutting tool and generally happens during interrupted cuts.

Thermal cracking is caused by very large rapid temperature changes at the cutting edge. Thermal cracks occur at right angles to the cutting edge of a tool and generally result during interrupted cutting or milling operations.

Fracture (Fig. 10-13E) occurs when a section of the tool insert breaks off because the forces of the cut are greater than the strength of the insert.

When *hardened steels* are cut, crater wear occurs behind the cutting-tool edge as the chip slides over the top of the tool. This crater wear tends to increase at cutting speeds above 400 ft/min (122 m/min). *Soft steels*, which produce a more ductile chip, increase the rate of crater wear and make such operations generally uneconomical for superabrasive tools.

Cast irons produce discontinuous chips and do not cause crater wear. The normal wear pattern in machining cast iron with PCBN is abrasive flank wear.

The only wear which should be taking place if the job is being done correctly is abrasive flank wear. Only flank wear is acceptable. Wear in any other place shows that the job is not being run correctly, set up correctly, or being run on the wrong workpiece. Check the machining guidelines in this text to make sure that each job is being run correctly.

USE OF COOLANTS

Light-duty, water-soluble oils of the type used for machining with carbide tools usually work well when machining with PCBN tools and inserts. The main

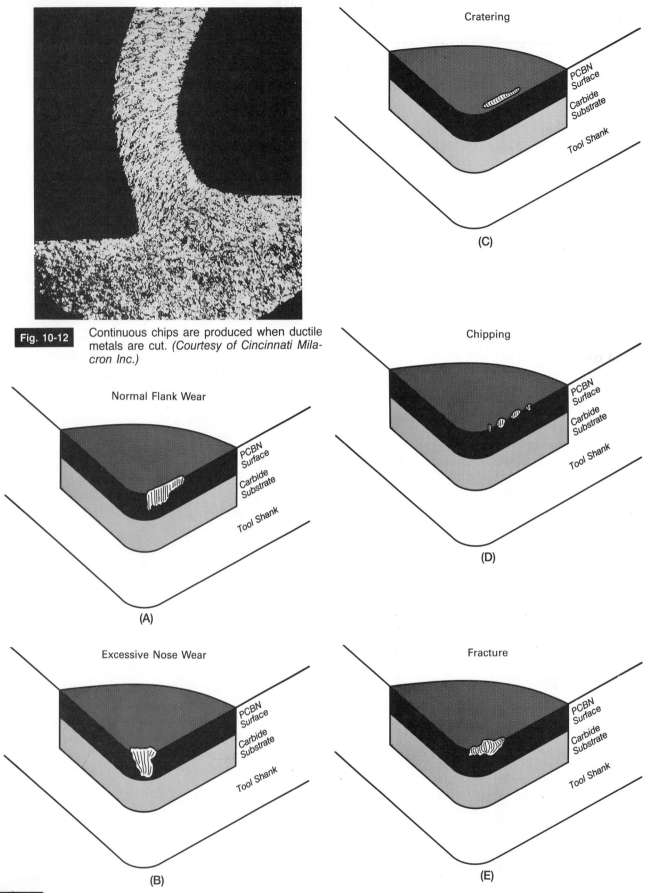

Fig. 10-12 Continuous chips are produced when ductile metals are cut. *(Courtesy of Cincinnati Milacron Inc.)*

Cratering

PCBN Surface

Carbide Substrate

Tool Shank

(C)

Normal Flank Wear

PCBN Surface

Carbide Substrate

Tool Shank

(A)

Chipping

PCBN Surface

Carbide Substrate

Tool Shank

(D)

Excessive Nose Wear

PCBN Surface

Carbide Substrate

Tool Shank

(B)

Fracture

PCBN Surface

Carbide Substrate

Tool Shank

(E)

Fig. 10-13 Types of cutting tool wear which may occur under certain conditions on PCBN tools. *(Courtesy of GE Superabrasives)*

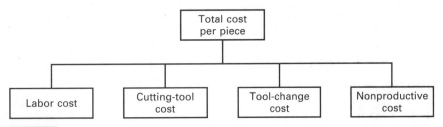

```
                    Total cost
                     per piece
                         |
     +---------------+---------+---------------+
     |               |         |               |
 Labor cost    Cutting-tool  Tool-change   Nonproductive
                  cost          cost            cost
```

Fig. 10-14   The factors which affect the total cost per piece of a manufactured part.

purpose of a coolant is to remove the heat from the cutting zone in order to protect the PCBN tool. This is very important when turning hardened steel in order to retard (slow) crater wear. Because of the unique properties of PCBN, lubrication is not as important as reducing the heat.

In some cases, it is desirable to machine with PCBN cutting tools using no coolant. In the particular case of milling with PCBN, the applications of any type of cutting fluid greatly increases the degree of hot and cold cycling (thermal cycling) of the PCBN cutting edges. Such extreme cycling can lead to rapid tool failure. *Always mill dry.*

There are other cases where a plentiful flow of cutting fluid should always be applied. These are:

- When taking higher depths of cut with tipped inserts or brazed-shank tools. The high conductivity of the PCBN layer could conduct enough heat to cause a braze failure or other damage to the tool shank.

- In grooving or other very deep cutting operations where coolants can assist in clearing the

chips and carrying heat away from the cutting zone.

- If the coolant system is not well maintained and/ or the coolant itself is poorly applied to the cutting zone. In such cases it could be just as well to turn the coolant off and machine the operation dry.

SPEEDS, FEEDS, AND DEPTH OF CUT

Operating conditions, such as cutting speed, feed rate, and depth of cut, control two very important variables in any machining operation. They are the *metal-removal rate* and the *cutting-tool life.* To achieve the lowest machining cost per piece and the highest production rates, the best operating conditions must be present to achieve the two variables. The total cost per piece is lowest at the cutting speed which is the best balance between the metal removal rate and the cutting tool life.

Cutting Speed

Cutting speed is the rate at which a point on the circumference of a workpiece passes the cutting tool in 1 min. It is generally measured in surface feet per minute (sf/min) or meters per minute (m/min). Cutting speed is the biggest factor which affects the life of a cutting tool and therefore has the greatest effect on the production rate and the total cost per piece.

Increasing cutting speed will reduce machining time, which, in turn, reduces machining costs. However, it may reduce cutting-tool life so that cutting tool and tool change costs increase.

- At very low cutting speeds, machining cost is very high because very few parts are being made; therefore, the total cost per piece is high.

- At very high cutting speeds, tool change costs are very high because frequent tool changes are required.

Cutting-tool cost is the cost of the insert divided by the number of pieces which can be effectively machined with it. *Tool-change cost* is the amount of labor, machine-tool cost, and overhead cost consumed in a tool change which can be charged to one piece. The

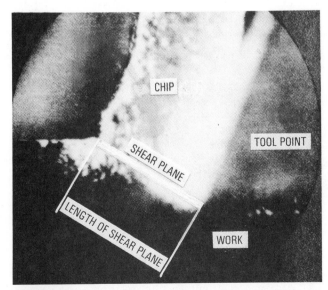

Fig. 10-15   During the cutting process, the metal is deformed along the shear plane producing heat. *(Courtesy of Cincinnati Milacron Inc.)*

Table 10-7 PCBN APPLICATION GUIDELINES

Classification	Material Types	Cutting Speed		Feed Rate	
		ft/min	m/min	in./rev	mm/rev
Hardened ferrous (>45 Rc)	Hardened steels (e.g., 4340, 8620, M-2, T-15)	250–350	76–106	0.006–0.020	0.15–0.50
	Hard cast irons (chilled iron, Ni-Hard)				
Superalloys (>35 Rc)	Nickel- and cobalt-base alloys (e.g., Inconnel, Rene, Stellite, Colmonoy)	650–800	198–244	0.006–0.010	0.15–0.25
Soft cast irons (typically 180–240 BHN)	Pearlitic gray iron, Ni-Resist	1500–3000	457–914	0.006–0.025	0.15–0.63
Flame-sprayed (>45 Rc)	Hard facing materials (e.g., Ni, Cr, WC)	200–350	60–106	0.006–0.012	0.15–0.30
Cold-sprayed materials (<45 Rc)	Hard facing materials (e.g., Ni, Cr, WC)	350–500	106–152	0.006–0.013	0.15–0.33

total cost per piece is the total of the labor cost, cutting-tool cost, tool change cost, and nonproductive cost (Fig. 10-14).

High cutting speed is the major factor in the amount of heat generated during a machining operation; however, it is one of the factors that make PCBN tools cut better. By increasing the cutting speed (revolutions per minute), hard materials can be cut in their softened state (above their yield temperature)

Fig. 10-16A Superabrasive inserts should be reground to the next inscribed circle (IC) for which there is a usable toolholder. *(Courtesy of GE Superabrasives)*

because the heat in the shear zone (Fig. 10-15) has softened the chip.

Table 10-7 gives recommended ranges of speeds and feeds for various types of metals. Since these are only general, it is wise to follow the cutting-tool manufacturer's recommendations for specific PCBN cutting tools and workpiece materials.

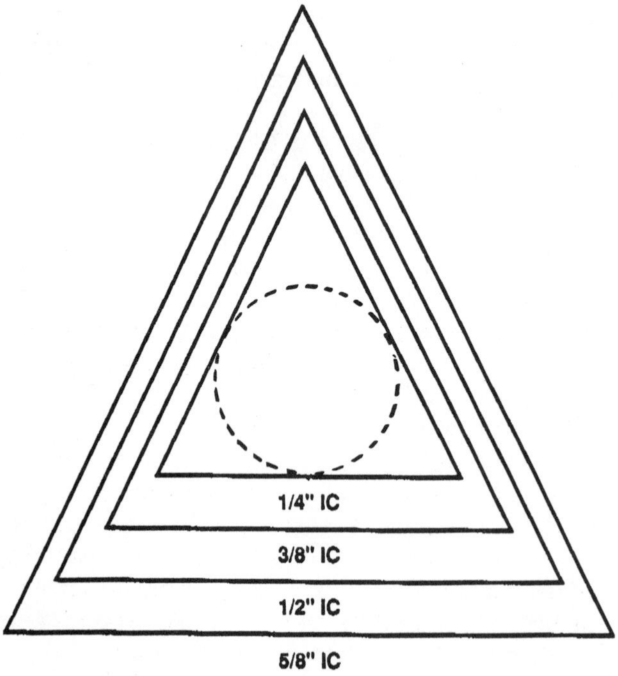

Feed Rate and Depth of Cut

Feed rate is the distance that a cutting tool moves along the length of a workpiece in one complete revolution. *Depth of cut* is the distance from the diameter of the work to the diameter of the section being cut measured at 90° to the work axis. Both the feed rate and the depth of cut will have a different effect on cutting-tool life.

The ideal cutting condition would be to rapidly produce heavy chips since 80 percent of the heat generated during machining is carried away by the chips. The remaining 20 percent of the heat will not cause any damage to the PCBN tool. A general rule to follow is to select the heaviest depth of cut and feed rate possible considering the rigidity of the machine and the horsepower available and the quality of the finished workpiece.

MACHINING PROCEDURE

The selection, setup, and operating conditions for PCBN tools must be just right if the machining operation is to succeed.

Speed

1. Always start within the recommended speed range for the type of material being machined and the grade of the PCBN tool or insert.

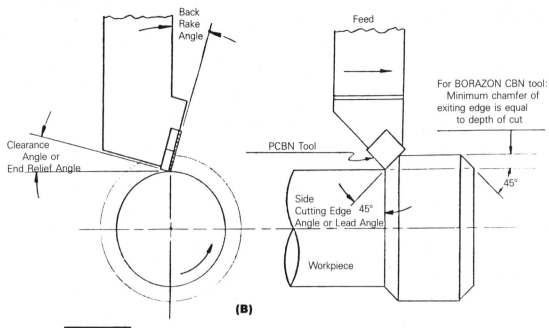

Fig. 10-16B PCBN tool setup for machining. *(Courtesy of GE Superabrasives)*

Feeds

2. Use the recommended feed rate wherever possible to obtain the best performance and longest life of the PCBN tool.
3. Use of feeds or depths of cuts less than 0.005 in. (0.12 mm) are not recommended.
   a. They produce a very light mass of chip which cannot carry away the heat of the cut.
   b. Excess heat at the cutting zone causes the work to expand, produces a tapered cut, and reduces the life of the tool.

Cutting-Tool Setup

4. Make sure that the pocket is clean and flat before installing an insert.
5. Do not clamp directly on the PCBN layer; use a chipbreaker or a suitable alternative to distribute the clamping forces. Clamp the insert securely; *do not* use a small pipe or tube on the Allen wrench.
6. Regrind the PCBN tool at the first sign of dullness.
   a. Both brazed-shank and tipped-insert tools can be reground. The amount of regrinding needed should be determined by the PCBN toolmaker.
   b. PCBN inserts can be reground to the smallest inscribed circle (IC) for which there may be a useable toolholder (Fig. 10-16A). This grinding must be done by a skilled PCBN toolmaker.
7. Never allow a wear land to grow into the carbide substrate. Heat, chatter, surface finish, and loss of workpiece accuracy will result.

8. Index and change tools on a regular basis.
   a. Dull tools will increase machining forces, causing chatter, which reduces the life of the PCBN tool.
   b. Impact damage can occur to dull tools more readily than to sharp tools during interrupted cuts.
9. Set the tool on center. If shims are required to bring the tool to the correct height, use only one shim of the correct thickness instead of a series of small shims.
10. Keep the overhang of PCBN tools as short as possible to prevent vibration and chatter.
11. Use negative-rake tools wherever possible.
12. Set the side cutting edge angle (SCEA) as close to 45° as possible (Fig. 10-16B). Avoid the use of a lead angle of less than 15°.
13. Use as large a nose radius on the PCBN tool as the job and machining operation will permit.

General Applications Guidelines

It is the aim of every company to produce the best-quality product at the lowest possible price to be competitive in the world marketplace. In order to do this, they must get the maximum production rates out of any machining operation. PCBN cutting tools make it possible to reach this goal on many applications; however, certain guidelines should be followed:

- Use PCBN tools for machining ferrous metals of Rc 45 hardness or higher and high-temperature alloys with Rc 35 or higher.
- Select a rigid machine with enough horsepower

to take advantage of the production potential PCBN cutting tools offer.

- Mount the cutting tool solidly and keep the tool overhang as short as possible to avoid deflection, chatter, and vibration (Fig. 10-22).

- Use negative-rake tools whenever possible (Fig. 10-8A).

- Choose the largest possible lead angle. Avoid lead angles of less than 15° (Fig. 10-8B).

- Wherever possible, use cutting fluids.

- Prepare the edge of the PCBN tool as outlined in Table 10-5 to suit the workpiece and the machining operation.

- When machining high-temperature alloys, always chamfer the exiting edge of the workpiece to reduce burr formation.

- Establish the life of each tool or cutting edge (usually after a certain number of pieces are cut) and change tools regularly.

- Always change tools when they become dull. Cutting with dull tools produces poor surface finishes and damages the surface quality.

- Establish speed and feed rates which result in a cost-effective combination of high productivity and long tool life. The speeds and feeds used when machining with carbides can be substantially increased with much longer tool life.

- Stop the cut as soon as chatter occurs; otherwise, the cutting tool will be damaged. Chatter indicates a dull cutting tool or a setup (work or cutting tool) which is not rigid.

Fig. 10-17 Turning a hardened Rc 69 steel roll with a PCBN cutting tool. *(Courtesy of GE Super-abrasives)*

MACHINING WITH PCBN TOOLS

Polycrystalline cubic boron nitride blank tools and inserts are widely used in the automotive, aerospace, and steel industries for machining hard and abrasive metals. Industry has found that these superabrasive cutting tools are among the most effective tools for production cost reduction and product improvement. Turning and milling operations on difficult-to-machine metals lend themselves to the use of PCBN cutting tools and in many cases eliminate the finishing operation of grinding.

Machining Hardened Steel

Polycrystalline cubic boron nitride cutting tools are widely used by industry throughout the world. They are used to machine hard, abrasive ferrous alloys, and difficult-to-machine nickel- and cobalt-base high-temperature alloys at high material-removal rates with long tool life.

Many of the materials that are routinely turned, bored, and milled with PCBN tools are so hard and

Table 10-8	HARDENED-STEEL ROLL

Machining operation—turning
        Material—modified 52100 steel, Rc 69
     Tool replaced—grinding

Machining Conditions

Depth of cut—0.045 in. (1.14 mm)
Cutting speed—275 sf/min (84 m/min)
Feed/revolution—0.022 in. (0.56 mm)
       Tool—BRNG 43 (round insert)
      Rake angle—5° (negative)
Clearance angle—5°
      Coolant—20:1 soluble oil

Economics—the previous grinding operation, which required 6 h, was reduced to 2 h of turning

abrasive that grinding was formerly the only practical conventional material-removal process. A few machining examples of the use of PCBN tools are as follows:

Fig. 10-18	Finishing a hardened gear with PCBN tools. *(Courtesy of Ingersoll Company)*

Example 1—Roll Turning

It was necessary to remove approximately 0.090 in. (2.28 mm) from the diameter of a roll 80 in. (2030 mm) long of modified 52100 steel hardened to Rc 69 (Fig. 10-17). The only method previously available was by grinding. A round PCBN insert was used for turning the roll to size, and the results are shown in Table 10-8 on page 175.

Example 2—Hardened Gear Milling

Hardened gears may be finished to size by lapping, grinding, machining, or a combination of these processes. Small gears are either lapped or ground, while large gears are usually machined and then ground to size and accuracy. Figure 10-18 shows an Ingersoll Company CNC gear-generating machine whose tungsten carbide inserts were replaced with PCBN inserts. The comparison results of this operation are shown in Table 10-9.

Table 10-9	HARDENED GEAR MILLING

Machine—Ingersoll gear milling machine
Operation—helical gear finishing
Material—4322 steel (Rc 60–65)
Tool replaced—tungsten carbide inserts

Machining Conditions	Tungsten Carbide	PCBN
Depth of cut	0.005 in. (0.12 mm)	0.005 in. (0.12 mm)
Cutting speed	175 ft/min (53 m/min)	325 ft/min (99 m/min)
Feed	0.004 in. (0.10 mm)	0.008 in. (0.20 mm)
Gears per cutter	1	10

Economics—Material removal rates increased (3 times)
    —Tool life improved (10 times)
    —Gear (pitch) accuracy increased

Fig. 10-19 The microstructure of pearlitic cast-iron engine block, which can be easily machined with PCBN tools. *(Courtesy of GE Superabrasives)*

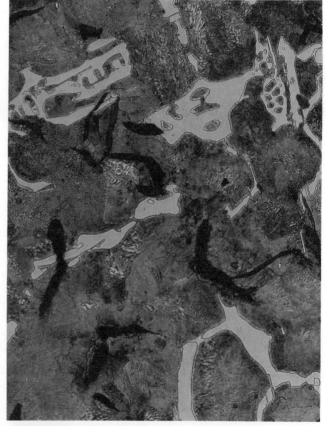

Fig. 10-20 The microstructure of nodular or ferritic cast iron which cannot be effectively machined with PCBN tools. *(Courtesy of GE Superabrasives)*

Machining of Cast Gray Iron with PCBN

Cast irons, just as most steels, are basically alloys of iron and carbon. By proper alloying, good control during manufacture, and various heat-treating operations, the properties of most cast irons can be varied greatly to suit specific purposes. This makes cast iron a very useful metal of great technical and economic importance.

The processes by which cast iron is made allow for a wide range of properties in the finished product. This, in turn, means that the structure of the iron, and in particular the exact way in which carbon and graphite are distributed in the structure, have a great effect on its properties.

Chapter 3 covered many of the factors which affect the machinability of metals. The specific metallurgical structure of cast iron affects its machinability, just as the structure of steel affects its machinability.

Gray cast iron with a fully pearlitic structure (Fig. 10-19) can be effectively machined with PCBN. In contrast to this, cast iron with a nodular or ferritic microstructure (Fig. 10-20) cannot be effectively ma-

Fig. 10-21 Machining a chilled cast-iron roll, Rc 61-62, with a PCBN cutting tool. *(Courtesy of GE Superabrasives)*

chined with PCBN. It is not completely understood why this is so; however, it is felt that it must relate to a chemical reaction which occurs between the PCBN and various types of cast iron during the machining process.

Example 3—Roll Turning

Chilled cast iron rolls with a hardness of Rc 61 to 62 needed occasional reconditioning which involved rough turning with tungsten carbide tools and the finish operation of grinding. Figure 10-21 shows this operation where the tungsten carbide tool inserts were replaced by PCBN inserts. The comparison results of this operation are shown in Table 10-10.

Table 10-10 CAST-IRON ROLL TURNING

Machining operation—turning
Material—chilled cast iron (Rc 61–62)
Tool and operation replaced—tungsten carbide and grinding

Machining Conditions	Tungsten Carbide	PCBN
Depth of cut	0.125 in. (3.17 mm)	0.100 in. (2.54 mm)
Cutting speed	50 sf/min (15.2 m/min)	300 sf/min (91.4 m/min)
Feed/revolution	0.020 in. (0.50 mm)	0.012–0.020 in. (0.30–0.50 mm)
Coolant	Compressed air	Compressed air
Cutting tool	Regular insert (1¼ × ¾ × ½ in.)	BRNG-43

Economics—Productivity more than doubled
　　　　　—Grinding operation eliminated

Example 4—Cast-Iron Milling

Many automotive parts such as engine blocks and cylinder heads (Table 10-11 on page 178) are still made of cast iron. PCBN inserts are being widely used for milling these components because of their long life and the high productivity they offer.

| Fig. 10-22 | Machining a hard, flame-sprayed valve seat with PCBN tools. *(Courtesy of GE Super-abrasives)* |

| Table 10-11 | **CAST-IRON CYLINDER HEAD** |

Machining operation—milling
Material—gray cast iron (190–250 BHN)
Tool replaced—silicon nitride (SiN)

Machining Conditions

Cutter—double negative (10 in. diameter/32 inserts)
Insert—SNG-62 (15° × 0.005 in. chamfer)
Depth of cut—0.020 in. (0.50 mm)
Cutting speed—3100 sf/min (945 m/min)
Feed/revolution—0.0047 in. (0.12 mm)
Coolant—dry

Economics—SiN produced 1900 pieces/corner
—PCBN produced 17,000 pieces/corner
—Less downtime for tool changes
—Decreased tooling setup time

Machining Hard Facings

Hard facing is a process for applying a thin weld deposit of special alloys on metal. The advantage of hard facing is that it can be applied to the exact area of a part where the most wear would occur to provide a hard, wear-resistant surface. Hard, flame-sprayed high-temperature alloys such as Colmonoy 43 with a hardness of Rc 55 can be readily machined with PCBN tools (Fig. 10-22). A comparison of turning flame-sprayed high-temperature alloy—Colmonoy 43—with tungsten carbide and PCBN inserts is shown in Table 10-12.

MACHINING PROBLEMS, CAUSES, AND REMEDIES

Polycrystalline cubic boron nitride cutting tools are designed primarily to machine hard, abrasive ferrous alloys and difficult-to-machine nickel- and cobalt-base high-temperature alloys. They are generally not recommended for superalloys lower than Rc 35 hardness, nor steels lower than Rc 45. When used properly, they are very cost-effective cutting tools which can produce high-quality parts.

Some of the most common problems, their causes and suggested remedies are outlined in Table 10-13.

Table 10-12 VALVE SEAT

Material—Colmonoy 43—flame-sprayed
Hardness—Rc 55
Operation—face, chamfer, bore

Machining Conditions	Tungsten Carbide	PCBN
Depth of cut	0.20–0.40 in. (0.050–1 mm)	0.040–0.080 in. (1–2 mm)
Cutting speed	142 ft/min (43 m/min)	285 ft/min (90 m/min)
Feed	0.004 in. (0.10 mm)	0.008 in. (0.20 mm)
Number of passes	2	1
Pieces/edge	3–4	1000

Economics—Tool life increased
　　　　　—Productivity increased

Table 10-13 PCBN TOOL PROBLEMS, CAUSES, AND REMEDIES

Problems	Causes	Remedies
Edge chipping	Improper edge preparation	Chamfer cutting edge (15° × 0.008) Ensure rigid toolholding system
Rapid tool flank wear	Cutting speed too slow 　—Insufficient to soften ahead of tool Cutting speed too fast 　—Excessive heat generated Feed rate too light 　—Thin chip can't dissipate heat 　—Tool rubbing Depth of cut too light 　—Excessive tool rubbing	Change speed to recommended rates 　—Hardened ferrous (>Rc 45) 　300–400 ft/min (91–122 m/min) 　—Soft gray cast iron (200 BHN) 　1500–3000 ft/min (457–914 m/min) Minimum feed rate (0.004 in. or 0.10 mm) Minimum depth of cut (0.005 in. or 0.12 mm)
Rapid tool crater wear	Tool soft steel Cutting speed too high 　—Excessive heat Insufficient coolant	Minimum hardness Rc 45 See above for speed recommendations Use coolant if possible

REVIEW QUESTIONS

Types and Sizes of PCBN Tools

1. Name the three categories of finished ready-to-use PCBN tools.
2. What type of inserts are generally more cost-effective? Explain why.

Properties of PCBN

3. Name the four main properties of PCBN cutting tools.
4. Compare the hardness and abrasion resistance of CBN with diamond and other abrasives.
5. What are PCBN composites, and where are they used?

PCBN Selection

6. Name the two main factors which affect the selection of PCBN tools.

Advantages of PCBN Cutting Tools

7. List four advantages that the tough, hard microstructure of PCBN tools offer the metalworking industry.
8. Briefly explain why the use of PCBN tools results in the following advantages:
   a. high-quality parts
   b. lower cost per piece

Machine-Tool Conditions

9. Name the five qualities that a machine tool should possess to operate at the high speeds and feeds necessary for PCBN tools.
10. Discuss the importance of a good coolant system.

Cutting-Tool Characteristics

11. Why is it wise to establish the life or cutoff point of a PCBN cutting edge?

12. What type of tool shape is
    a. the strongest?
    b. the most versatile?
13. Name the two points about tool geometry which help to obtain longest tool life, efficient metal removal rates, and good surface finishes.
14. What two methods can be used to overcome the problem of sharp corners on PCBN tools?
15. State three reasons why it is advisable to use as large a nose radius as possible.

Principles of Metal Cutting

16. What determines the type of chip which will be produced during a metal-cutting operation? Give an example of each.
17. Name the type of metal which produces the following chip types:
    a. continuous
    b. discontinuous

Use of Coolants

18. What is the main purpose of a coolant?

19. Why should coolant not be used for milling operations?

Speeds, Feeds, and Depth of Cut

20. Name the two important variables in any machining operation which are controlled by the cutting speed, feed rate, and depth of cut.
21. Define:
    a. cutting-tool cost
    b. tool-change cost
    c. total cost per piece
22. Which has more effect on the cutting life: feed rate or depth of cut?

Machining Procedure

23. Why is it not advisable to have a feed or depth of cut less than 0.005 in. (0.12 mm)?
24. List four important points which should be observed regarding the setup of the PCBN tool.

CHAPTER ELEVEN

Polycrystalline Diamond Tools

The introduction of polycrystalline diamond (PCD) tool blanks by General Electric Company (GE) in 1973 began a new age in the machining of nonferrous metals and abrasive nonmetallic materials. The ability to bond a layer of diamond abrasive to a cemented carbide substrate produced a cutting-tool blank of extraordinary hardness, strength, and abrasion resistance. This chapter will explain polycrystalline diamond (PCD) tools and how they are used.

OBJECTIVES

After completing this chapter you should be able to:

1. Understand the properties of PCD tools and how they differ from other cutting tools
2. Select the proper type and size of PCD cutting tools for machining various materials
3. Set up the cutting tool and machine for cutting with PCD tools
4. Understand the conditions necessary to achieve maximum productivity

TYPES AND SIZES OF PCD TOOLS

Polycrystalline diamond tool blanks are a combination of a layer of diamond integrally bonded to a cemented tungsten carbide substrate. The blanks are produced in a high-pressure high-temperature apparatus as cylindrical disks and then cut into segments, triangles, rectangles, or other special shapes by wire-cut electrical discharge machining. Table 11-1 lists the combination of shapes and grades which are available to tool manufacturers for the production of PCD cutting tools.

Polycrystalline diamond is produced in two major categories—carbide-supported blanks and unsupported thermally stable PCD. Only the carbide-supported blanks used in cutting-tool applications are discussed in this chapter. Thermally stable PCD is used only in drilling applications in the oil, gas, and mining industries and in grinding-wheel dressing and truing applications. The carbide-supported PCD is available in three microstructures to suit various machining applications. The basic difference among the three series is the size range of the diamond particles which are used to manufacture the polycrystalline blank. The blank series and their description are listed as follows:

1. *Fine PCD Blanks* (1600 series in Table 11-1) are manufactured from fine diamond particles of approximately 5 μm diameter. This fine-grain structure allows for the development of extremely sharp cutting edges. These PCD blanks have been developed principally for woodworking and are not used extensively in metalworking.
2. *Medium PCD Blanks* (1300 series in Table 11-1) are composed of fine to medium-fine crystals having an average diameter of 7 μm. This is a widely used type of PCD blank for machining nonferrous and nonmetallic materials.
3. *Coarse PCD Blanks* (1500 series in Table 11-1) are composed of diamond particles having an average diameter of 25 μm. They are used for machining cast aluminum alloys, especially those containing more than 16 percent silicon. The 1500 series PCD tools have superior impact resistance and are suitable for interrupted turning and milling applications.

The diamond microstructure plays a big part in determining the characteristics of the PCD tool blank and its applications, tool life, finish, etc. A comparison of the abrasion resistance of the three microstructures is shown in Fig. 11-1A. Some typical surface

Table 11-1 COMPAX DIAMOND TOOL BLANKS

SEGMENTS

Product Number	Nominal Dimensions mm			
	ℓ°	d	ℓ	t
1311	45	—	3.8	1.5
1315	60	—	3.8	3.2
1316	60	—	3.8	1.5
1320	90	—	3.9	3.2
1321	90	—	3.9	1.5
1325	180	—	8.1	3.2
1326	180	—	8.1	1.5
1330	360	8.1	—	3.2
1331	360	8.1	—	1.5
1511	45	—	6.3	1.5
1515	60	—	6.4	3.2
1516	60	—	6.4	1.5
1520	90	—	6.5	3.2
1521	90	—	6.5	1.5
1525	180	—	13.2	3.2
1526	180	—	13.2	1.5
1530	360	13.2	—	3.2
1531	360	13.2	—	1.5
1611	45	—	6.3	1.5
1615	60	—	6.4	3.2
1616	60	—	6.4	1.5
1620	90	—	6.5	3.2
1621	90	—	6.5	1.5
1625	180	—	13.2	3.2
1626	180	—	13.2	1.5
1630	360	13.2	—	3.2
1631	360	13.2	—	1.5

TRIANGLES

Product Number	Nominal Dimensions mm				
	ℓ°	d	ℓ	t	w
60T5/1.5 1300	60	—	5.0	1.5	—
90T5/1.5 1300	90	—	5.0	1.5	—
60T5/1.5 1500	60	—	5.0	1.5	—
90T5/1.5 1500	90	—	5.0	1.5	—
90T7.5/1 1500	90	—	7.5	1.5	—

RECTANGLES

	ℓ°	d	ℓ	t	w
1566	90	—	8.0	3.2	2.0
1575	90	—	5.0	3.2	2.5
1577	90	—	11.5	3.2	2.5

34mm ROUNDS

	ℓ°	d	ℓ	t	w
360R34/3 1300	360	34.0	—	3.2	—
360R34/3 1500	360	34.0	—	3.2	—
360R34/3 1600	360	34.0	—	3.2	—

(Nominal diamond layer thickness is 0.5 mm)

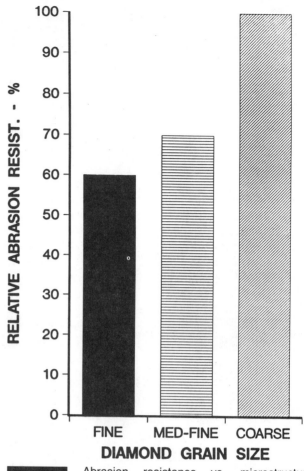

Fig. 11-1A Abrasion resistance vs. microstructure. *(Courtesy of GE Superabrasives)*

finish values obtained in machining with PCD are compared in Fig. 11-1B. PCD microstructure is only one of many factors which will influence surface finish.

Polycrystalline diamond tool blanks are available from tool suppliers in a wide variety of shapes and sizes to suit various machining applications. PCD cutting tools are available as brazed tipped inserts and brazed shank tools. See Table 11-2 for the recommended application guidelines for PCD tooling.

PROPERTIES OF PCD TOOLS

Selected properties of PCD tool blanks are compared to other cutting-tool materials in Tables 9-5A, 9-5B, and Fig. 9-6. The *cemented tungsten carbide substrate* provides an excellent mechanical support for the polycrystalline diamond layer and imparts a toughness to the finished PCD tool.

The important properties of the *diamond layer* of PCD tools are hardness, abrasion resistance, compressive strength, and thermal conductivity. The dia-

mond crystals in the polycrystalline superabrasive layer of a PCD tool blank are randomly oriented. The hardness and abrasion resistance of the diamond layer are, therefore, uniform in all directions. Unlike single-crystal mined diamond, there are no hard or soft planes, or weak-bound planes that can lead to gross cleavage (breaking of the crystal). Therefore, there is no need for special orientation (placement) of the diamond tool blank to achieve the best machining results.

The *compressive strength* of the diamond layer is the highest of all cutting tools. This is due to its dense structure, which enables PCD cutting tools to withstand the forces created during high material-removal rates and the shock of interrupted cuts. The *thermal conductivity* of the PCD layer is the highest of any cutting tool. This allows greater dissipation (transfer) of the heat created at the chip-tool interface, especially when cutting tough, abrasive materials at high metal-removal rates.

ADVANTAGES OF PCD CUTTING TOOLS

The advantages that PCD cutting tools offer industry more than offset their initial higher costs. PCD tools

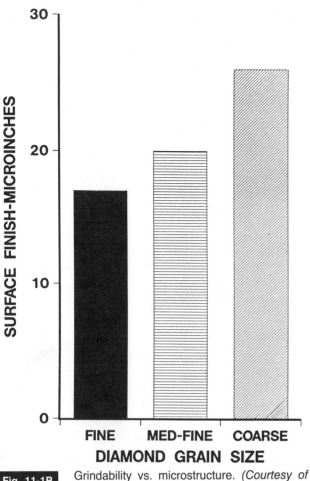

Fig. 11-1B Grindability vs. microstructure. (Courtesy of GE Superabrasives)

Table 11-3	POLYCRYSTALLINE DIAMOND TOOLING APPLICATIONS
Nonferrous Metals	**Nonmetallic Materials**
Silicon aluminum alloys	Epoxy resins
Brass and bronze alloys	Fiberglass composites
Copper alloys	Carbon phenolics
Magnesium alloys	Hard rubber
Zinc alloys	Plastics
Presintered tungsten carbide	Ceramics
Sintered tungsten carbide	Wood-fiber products

are capable of greatly improving efficiency, reducing scrap parts, and increasing the quality of the product when used to machine nonferrous and nonmetallic materials (Table 11-3). Some of the main advantages of polycrystalline diamond cutting tools are shown in Fig. 11-2.

Long Tool Life. Because of their high uniform hardness and wear resistance, PCD cutting tools resist chipping or cracking; therefore, their tool life is generally predictable. The rate of wear at the cutting edge is far slower than that of cemented tungsten carbide tools. Reduced tool wear results in holding closer tolerances on workpieces, requiring fewer tool adjustments.

Cutting Tough, Abrasive Material. PCD cutting tools are designed to machine tough, abrasive nonferrous and nonmetallic materials faster and at a lower cost than cemented tungsten carbide or single-crystal mined-diamond tools. The sharp cutting edge shears the chip cleanly and reduces the friction force of the chip sliding over the rake surface of the tool.

High-Quality Parts. The sharpness of PCD blank tools results in an efficient cutting action which consistently produces better part accuracy and geometry and results in less scrap being produced. The need for part inspection is greatly reduced.

Fine Surface Finishes. PCD cutting-tool blanks are polished and ground to obtain a very keen cutting edge which produces a very effective cutting action. These tools produce consistently good surface finishes throughout a production run without cutting edge wear or chipping, which would affect the surface finish. PCD cutting tools do not produce the ultrafine surface finishes possible with single-point mined diamond tools; however, a finish as fine as 5 μin. has been obtained under special conditions.

Reduced Machine Downtime. Because PCD tools are very hard and maintain a keen cutting edge for long production runs, there is less time required to

Table 11-2	PCD BLANK APPLICATION GUIDELINES
PCD grade	**Application**
Medium grain (1300 series)	General-purpose in all nonferrous metals
Coarse grain (1500 series)	Milling, interrupted cuts
Fine grain (1600 series)	Woodworking and fine finish requirements

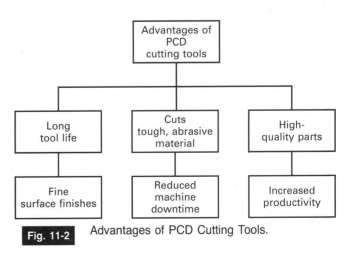

Fig. 11-2 Advantages of PCD Cutting Tools.

index, change, or recondition the cutting tool. This also eliminates the need to adjust tool positions and, in many cases, the need for compensation devices. This results in less machine downtime, which reduces the machining time for each part.

Increased Productivity. A combination of all the advantages that PCD tools offer, such as increased speeds and feeds, long tool life, longer production runs, improved size control, consistent part quality, and savings in labor costs, all have an effect on overall production rates and the manufacturing cost per piece.

MACHINE-TOOL REQUIREMENTS

Machine tools used to produce parts with cemented carbide cutting tools are generally suitable for machining with PCD tools.

Machines should be inspected for loose bearings, gibs, or fixtures and unsatisfactory conditions corrected to avoid poor cutting action, which shortens the life of the cutting edge.

Newer machine tools are designed to increase the effectiveness of PCD tool blanks by using higher cutting speeds for machining than those used for tungsten carbide cutting tools. The results are more pieces per hour, a high-quality workpiece finish, more consistency in dimensional accuracy, less scrap, and less downtime due to the long life of the cutting edge.

GENERAL GUIDELINES FOR MACHINING WITH PCD

Polycrystalline diamond tools must be properly prepared and used under the conditions which will provide the highest machining performance. To maintain good control over part quality and finish, it is always wise to establish the life or cutoff point of a cutting edge (usually after a certain number of pieces are cut) and make a practice of changing cutting tools at this point. If PCD cutting tools are run until they are excessively worn, they will cut inefficiently and increase machining temperatures and the cutting force required. This will, in turn, produce poor surface finishes and affect part accuracy and geometry.

Table 11-4 provides machining condition guidelines for several important workpiece materials. These guidelines represent the range of conditions within which PCD is being effectively used to machine each material. It is wise to always start at the lowest speeds and feeds in each range. First establish that PCD is producing fully satisfactory parts which meet all specifications; only then should increases in operating conditions be investigated to take advantage of PCD productivity potential.

Tool Geometry

The shape of the cutting tool and how it meets the workpiece are very important factors which determine how well and how long it will cut efficiently. To obtain the longest tool life, efficient metal-removal rates, and acceptable surface finishes, the following points should be observed.

A good starting point for a PCD cutting tool is to use the same tool geometry (shape) as the cemented tungsten carbide tool which it replaced. This shape can be modified through experience to provide a balance between the lowest total machining cost per piece and the longest cutting-tool life.

To avoid a weak support for the cutting edge, it is recommended that relief angles be as small as possible and not more than 15°. Increasing the front or side relief angles beyond 15° could result in greatly reducing the firm base that supports the polycrystalline cutting edge and the destruction of a valuable tool.

Polished Rake Surface

Virtually all machining of metals with PCD require that the top rake surface be polished. This polishing operation, performed by the tool manufacturer, provides much smoother chip flow, a sharper cutting edge, and reduces the risk of metal buildup on the cutting edge. *Always machine metals with a polished rake surface on the PCD tool.*

CUTTING FLUID SELECTION

Polycrystalline diamond blank tools can be used to machine parts either wet or dry. In most cases, however, tool performance is improved by the use of a *properly applied* cutting fluid. Soluble oil-water emulsions similar to those used when machining with cemented tungsten carbide tools are widely used with excellent results. Their lubricating qualities help to reduce frictional heating and the formation of built-up edges while providing good chip flow.

To be effective, a cutting fluid must be supplied in a large steady flow to the rake surface of the PCD tool. Certain applications of PCD tools are used without any type of cutting fluid. The high thermal conductivity of PCD allow the tools to be used dry for many applications without reducing tool life.

SPEEDS, FEEDS, AND DEPTH OF CUT

Operating conditions, such as cutting speed, feed rate, and depth of cut, control two very important variables in any machining operation: the *metal-removal rate* and the *cutting-tool life*. To achieve the lowest machining cost per piece and the highest production rates, the best operating conditions must be present to achieve the two variables. The total cost

Table 11-4   PCD TOOL STARTING CONDITIONS

Material	Cutting Speed		Feed Rate		Back Rake Angle	Side Rake Angle
	ft/min	m/min	in./rev	mm/rev		
Aluminum alloys						
(4–8% Si)	4200–6500	1280–1980	0.004–0.025	0.1–0.63	0 to 5°	0 to 8°
(9–14% Si)	3300–5200	1000–1580	0.004–0.020	0.1–0.50	0 to 5°	0 to 8°
(16–18% Si)	1000–2300	305–700	0.004–0.015	0.1–0.40	0 to 5°	0 to 8°
Copper alloys	900–3000	300–1000	0.002–0.006	0.05–0.15	10 to 20°	15 to 20°
Tungsten carbide	45–150	15–50	0.004–0.012	0.1–0.30	−5°	0 to −5°
Manufactured wood	6000–13000	2000–4500	0.002–0.006	0.5–0.15	5 to 10°	0 to 8°
Graphite epoxy composites	400–1500	120–450	0.002–0.006	0.5–0.15	5 to 10°	0 to 5°

per piece is lowest at the cutting speed which is the best balance between the material-removal rate and the cutting-tool life.

Cutting Speed

Cutting speed is the biggest factor which affects the life of a cutting tool and therefore has the greatest effect on the production rate and the total cost per piece. Increasing the cutting speed will reduce machining time, which, in turn, reduces machining costs. However, it may reduce cutting-tool life so that cutting-tool and tool-change costs increase.

- At very low cutting speeds, machining cost is very high because very few parts are being made; therefore, the total cost per piece is high.

- At very high cutting speeds, tool change costs are high because tool life is reduced and more frequent tool changes are required.

Cutting-tool cost is the cost of the insert or cutting tool divided by the number of pieces which can be effectively machined with it. *Tool-change cost* is the amount of labor, machine-tool cost, and overhead cost consumed in a tool change which can be charged to one piece. The *total cost per piece* is the total of the labor cost, cutting tool cost, tool change cost, and nonproductive cost.

Polycrystalline diamond tools follow the same laws of tool wear that cemented carbide tools must follow. The higher the work speed, the greater the tool wear. The main difference is that the rate of tool wear with PCD is much lower as speed increases in comparison to carbide. The tool life chart in Fig. 11-3 shows a comparison of the effect which speed has on a typical carbide tool and a PCD tool in machining a high-silicon-aluminum alloy. For example, if a work speed of 800 ft/min (244 m/min) were selected, carbide tool life would be 3 min, while PCD life would be well over 100 min and off this chart.

Good judgment should always be used in selecting the best combination of speed and feed for a given application. Refer to Table 11-4 for recommended cutting speeds, feeds, and tool specifications for various types of materials. It is also wise to follow the cutting-tool manufacturer's recommendations for specific PCD cutting tools and workpiece materials.

Feed Rate and Depth of Cut

Feed rate is the distance that a cutting tool moves along the length of a workpiece in one complete revolution. *Depth of cut* is the distance from the diameter of the work to the diameter of the section being cut measured at 90° to the work axis. Feed rate and depth of cut can be maximized, but close attention must be given to tool geometry (shape). When using high feed rates and depths of cut, relief, back rake, and side rake should be less than in the case of carbide tools at the same feeds and depths. A general rule to follow is to select the heaviest depth of cut and feed rate possible considering the rigidity of the machine and the horsepower available.

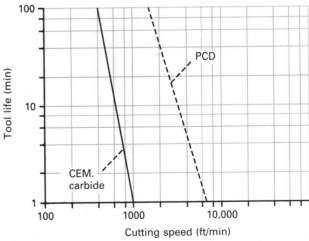

Fig. 11-3   Relationship between cutting speed and tool life.

MACHINING EXAMPLES

Polycrystalline diamond tool blanks and inserts are widely used in the automotive, aerospace, and other manufacturing industries for turning and milling operations for abrasive nonferrous and nonmetallic materials. Industry has found that these superabrasive cutting tools are among the most effective tools for production cost reduction and part quality improvement. In terms of the number of pieces per cutting edge, downtime, and overall productivity, PCD tools have proved to be the most cost-efficient tools available today.

General Applications Guidelines

To obtain the best tool performance and the most number of parts per cutting edge, the following guidelines should be closely followed:

- Use PCD cutting tools to machine only nonferrous and nonmetallic materials.

- Select a rigid machine with enough horsepower to maintain the cutting speed where PCD tools perform best.

- PCD tools are very effective even when run at conventional carbide speeds.

 NOTE The maximum productivity of PCD can be obtained by operating at higher speeds.

- Establish speed and feed rates which will result in a cost-effective combination of high productivity and long cutting-tool life.

- Use rigid toolholders and keep the tool overhang as short as possible.

- Generally, the same tool geometry as that used for tungsten carbide tools is satisfactory.

- Use positive-rake angles whenever possible.

- Use the largest nose radius possible for better surface finishes and to spread the cutting force over a wider area.

- Use PCD tools with polished rake faces to reduce the friction of the chip and to produce better surface finishes.

- Establish the life of each cutting edge or tool (usually after a certain number of pieces are cut) and change tools regularly.

- Use coolant wherever possible to reduce heat, promote free cutting, and flush away the abrasive chips from the finished work surface.

Table 11-5 **MACHINING COMPARISONS BETWEEN CEMENTED CARBIDE AND PCD TOOLS**

Material		Cemented Carbide Tool	PCD Tool
Silicon aluminum alloy pistons (turning)		Initial tool cost = X 700–1200 pieces/cutting edge 3 cutting edges/tool 2100–3800 pieces/tool Unacceptable surface finish	Initial tool cost = 10X 50,000 pieces/cutting edge 1 cutting edge/tool 50,000 pieces/tool 18 μin.
Silicon aluminum alloy pistons (boring)		Initial tool cost = X 1200 pieces/tool 720 pieces/hour Tool cost/piece = X/1200	Initial tool cost = 24.5X 200,000 pieces/tool 900 pieces/hour Tool cost/piece = X/8100
Silicon aluminum intake manifolds (milling)		Tool cost/piece = X Cutting edges/insert = 4 Pieces/cutting edge = 6200 Number of regrinds = 0 Pieces/insert = 24,800	Tool cost/piece = 1.5X Cutting edges/insert = 1 Pieces/cutting edge = 89,100 Number of regrinds = 3 Pieces/insert = 356,400
Metal-filled abrasive rubber rings (turning)		Cutting edges/insert = 4 Pieces/cutting edge = 35 Pieces/insert = 140 Tool life = X Acceptable surface finish	Cutting edges/insert = 1 Pieces/cutting edge = 6000 Pieces/insert = 6000 Tool life = 171X Improved surface finish
Copper alternator slip rings (turning)		Initial tool cost = X Cutting edges/insert = 3 Pieces/cutting edge = 30 Pieces/insert = 90 Surface finish = 70–75	Initial tool cost = 18X Cutting edges/insert = 1 Pieces/cutting edge = 10,000 Pieces/insert = 10,000 Surface finish = 40–45
Abrasive glass-filled polypropylene		Tool life = X Parts/tool = 100–200 Parts/hour = 70 Scrap rate = 30%	Tool life = 100X Parts/tool = 40,000 Parts/hour = 85 Scrap rate = 5% approximately

These guidelines are based on experience gained in thousands of successful production applications. PCD tools generally outperform conventional diamond tools and cemented tungsten carbide tools by wide margins. When a cutting tool is performing well, it should be studied with a view to changing tool geometry or machining conditions in order to achieve even better tool performance.

MACHINING OPERATIONS

Some specific examples of the practical uses of PCD in industry are found in Table 11-5. Each example shows how the properties of PCD can be used to increase production, and in most cases, part quality as well.

REVIEW QUESTIONS

1. Why is polycrystalline diamond (PCD) considered to be a highly efficient cutting tool?

Types and Sizes of PCD Tools

2. Describe the composition of a PCD tool blank.
3. Name the three series of PCD tool blanks.
4. What series of PCD tool should be used for
   a. rough turning?
   b. producing fine surface finishes?
5. How does the PCD tool blank's diamond grain size affect
   a. abrasion resistance?
   b. surface finish?

Properties of PCD Tools

6. List the four main properties of PCD tools.
7. Why is thermal conductivity important to a cutting tool?
8. List six advantages of using PCD tools for machining operations.

Machine-Tool Requirements

9. What five characteristics must a machine tool have in order to use PCD tools?

General Guidelines

10. Why is it important to establish the life or cutoff point of a cutting edge?
11. How does the tool geometry of a PCD tool compare to that of a tungsten carbide tool?

Cutting Fluid Selection

12. List the advantages of properly applied cutting fluids when machining with PCD tools.

Speeds, Feeds, and Depth of Cut

13. What two variables are controlled by the speed, feed, and depth of cut?
14. Discuss the effects of:
    a. very low cutting speeds
    b. very high cutting speeds.
15. What is the general rule for speeds and feeds when machining with PCD tools?

Machining Examples

16. Briefly describe the effects of:
    a. establishing good speed and feed rates
    b. rigid toolholders and short tool overhang
    c. large nose radius
    d. polished rake face

GLOSSARY OF TERMS

amorphous carbon—carbon in which there is no uniform arrangement of carbon atoms as in the case of graphite or diamond

arc of contact—the length of wheel circumference which is in direct contact with the workpiece during grinding

ASTM—abbreviation for American Society for Testing Materials

belt apparatus—a term describing the punch-and-die system designed to withstand the high pressure and high temperature required to manufacture superabrasives

Borazon CBN—trademark of the General Electric Company for its cubic boron nitride (PCBN) abrasive products

BZN tools blanks—a trademark of the General Electric Company for its polycrystalline cubic boron nitride (PCBN) tool blanks

carbon solubility potential—a characteristic of ferrous metals which describes their potential for further reaction with carbon

CBN—abbreviation for cubic boron nitride

cemented tungsten carbide—a cutting-tool material manufactured by sintering finely divided carburized tungsten metal powder in a cobalt metal binder

chip-tool interface—the area defining the contact between the face of a cutting tool and the chip being removed from the workpiece

cleavage—the fracturing of an abrasive grain along one or more of its principal planes

Compact—a generic name for sintered polycrystalline superabrasives

Compax blanks—a trademark of the General Electric Company for its polycrystalline diamond (PCD) tool blanks

composite—a material created by combining two or more materials; often a reinforcing element and a compatible binder

concentration—a term used for the amount of superabrasive contained per unit volume of grinding-wheel bond material

concentricity—the condition in which the periphery of a grinding wheel and its geometric center are on a common axis

conventional abrasives—a term referring to aluminum oxide or silicon carbide abrasives

core—the center or body of the grinding wheel to which the superabrasive is bonded

cost-effective—a description of a grinding or machining operation where the increase in productivity and quality through the use of superabrasives more than justifies their cost

cost per part—the total cost of producing a part when all costs such as superabrasive tools, materials, labor, and overhead are included

crater wear—an undesirable form of cutting-tool wear which occurs just behind the cutting edge on the top of the tool due to friction from the workpiece chip sliding over the top of the tool

creep-feed grinding—a method of grinding in which the table speeds are very low and the wheel is fed down to full depth of cut in one or two passes

crush dressing—a method of creating a specific form in a specially bonded superabrasive wheel by feeding the wheel at slow speed into a carbide or hardened steel roll, which is a reverse of the desired form

CGS-II—trademark of the General Electric Company for a diamond abrasive designed for grinding composite cemented carbide and steel

cubic boron nitride—a material harder than any other material except diamond; manufactured under high-pressure high-temperature conditions using hexagonal boron nitride

dressing—resharpening and renewing the cutting surface of a superabrasive wheel by relieving the bond from around the abrasive grains

DTG—abbreviation for "difficult to grind"

electroplated bond—a process of holding superabrasives to the periphery of a grinding wheel with an electrodeposited nickel bond

ETG—abbreviation for "easy to grind"

FEPA—abbreviation for European Federation of Abrasive Manufacturers

flank wear—the normal type of cutting tool wear in which the flank of a cutting tool wears because of the sliding contact with the workpiece

friability—that property of an abrasive grain which describes its tendency to fracture during the impact of grinding

frictional heat—the heat created when a wheel or cutting tool rubs the workpiece surface

functional costing—a new costing technique which is used to justify large expenditures on new tools and major manufacturing equipment

grindability—the relative ease by which a metal can be ground

grinding ratio (G ratio)—the ratio of the volume of work material removed to the volume of wheel worn during grinding **under** a specific set of conditions

grit size—that characteristic of a grinding wheel which the manufacturer uses to designate the nominal size of the abrasive contained in the wheel

hexagonal boron nitride—a white, powdery substance, sometimes known as *white graphite*, used in the manufacture of cubic boron nitride

impregnated bond—a method of attachment of superabrasive grains to the rim or face of a grinding wheel by uniformly distributing the grains through the thickness of the bond material

machinability—the relative ease by which a metal can be machined with a cutting tool

macrofracture—a type of wear of an abrasive grain where cleavage takes place along a principal plane

Man-Made Diamond—a trademark of the General Electric Company for its manufactured diamond abrasive products

martensite—a structure or metallurgical phase in steel developed by quenching after heating; required for high hardness and strength

martensitic steel—a steel with carbon content and alloying elements capable of forming martensite when heated and quenched

mesh size—the size of a superabrasive grain as defined by an internationally approved sieving method

metal bond—a form of impregnated bond in which sintered powdered metal is used to bond the superabrasive to the grinding wheel

microcrystalline—same as polycrystalline (see **polycrystalline**) except that individual crystallites are of micron or submicrometer in size

microfracture—a type of superabrasive grain wear in which extremely small particles break from the surface of the abrasive grain in the grinding process

monocrystalline—a single crystal with a regular arrangement of atoms in all three dimensions

MRR—abbreviation for material-removal rate; the time rate at which material is removed from a workpiece measured in cubic inches or cubic centimeters per unit time

out-of-round (out-of-truth)—a condition where the rim or face of the grinding wheel is not concentric (true) with the spindle axis of rotation

PCBN—abbreviation for polycrystalline cubic boron nitride

PCD—abbreviation for polycrystalline diamond

peripheral grinding—the process of grinding with the outside diameter or periphery of a grinding wheel

plastic deformation—a permanent change in dimensions or set which occur in a metal body as a result of excessive tensile, shear, or compressive stresses

polycrystalline—a crystalline body composed of many small crystals randomly or partially randomly oriented; individual crystals are large enough to be identified under a microscope at $20 \times$ magnification

productivity—the efficiency or effectiveness with which tools, raw materials, and labor can be applied in the production process

residual stress—stresses present in a free metal body, usually as a result of some prior treatment such as grinding, severe temperature changes during quenching, or chemical differences as in carburized surfaces

resin bond—a form of impregnated bonding in which a hot-pressed phenolic resin is used to bond the superabrasive to the rim of the grinding wheel

ROI—abbreviation for "return on investment"

runout—see **out-of-round** or **out-of-truth**

RVG—a trademark of the General Electric Company for a diamond abrasive suited primarily for use in resinoid and vitrified grinding wheels

RVG-W—a trademark of the General Electric Company for a metal-coated diamond abrasive suited specifically for use in resinoid grinding wheels

RVG-D—a trademark of the General Electric Company for a copper metal-coated diamond abrasive suited specifically for use in resinoid bonded grinding wheels for dry-grinding cemented carbides

self sharpening—a wear characteristic of an abrasive grain which describes its ability to continually develop sharp cutting points through a process of micro- or macrofracturing

shear properties—characteristics of metal deformation in which parallel planes within the metal are displaced by sliding but keep their parallel relation to each other

sparkout—a grinding pass taken over a surface with no downfeed to eliminate previous grinder spindle or wheel deflection

substrate—the base of cemented carbide which provides support and strength for the layer of polycrystalline superabrasive bonded to its surface

superabrasive—a term referring to diamond and cubic boron nitride abrasive grains and polycrystalline diamond and cubic boron nitride cutting-tool blanks

superalloys—alloys developed for very high temperature work where there are relatively high stresses and where oxidation resistance is required

swarf—grinding debris consisting of metal chips ground from the workpiece and abrasive particles worn from the grinding wheel

thermal conductivity—that characteristic of a material which defines its capability to transfer heat

tool-work interface—the area where the edge of a cutting tool and the workpiece contact

truing—shaping the cutting surface of a grinding wheel to the required form and bringing it into concentricity (making it true) with the axis of rotation

vitrified bond—a form of impregnated bond in which the superabrasive grains are attached to the rim of the grinding wheel by the sintering of glassy materials

weight percent (wt %)—the percentage of the total weight of a metal-coated superabrasive made up by the coating itself

work harden—an increase in strength and hardness of certain metals as they are machined or formed

work wheel interface—see **tool-work interface**

INDEX

191